MORE METRICS

with scissors,
ta...
S
S

You dialed 14.7 cm!

That's 147 mm!

.147 meters!

SCIENCE WITH SIMPLE THINGS SERIES

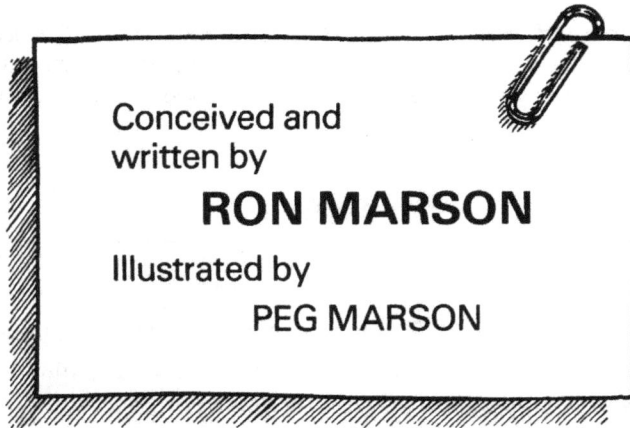

Conceived and
written by

RON MARSON

Illustrated by

PEG MARSON

TOPS LEARNING SYSTEMS

342 S Plumas Street
Willows, CA 95988

www.topscience.org

WHAT CAN YOU COPY?

Dear Educator,

Please honor our copyright restrictions. We offer liberal options and guidelines below with the intention of balancing your needs with ours. When you buy these labs and use them for your own teaching, you sustain our work. If you "loan" or circulate copies to others without compensating TOPS, you squeeze us financially, and make it harder for our small non-profit to survive. Our well-being rests in your hands. Please help us keep our low-cost, creative lessons available to students everywhere. Thank you!

PURCHASE, ROYALTY and LICENSE OPTIONS

TEACHERS, HOMESCHOOLERS, LIBRARIES:

We do all we can to keep our prices low. Like any business, we have ongoing expenses to meet. We trust our users to observe the terms of our copyright restrictions. While we prefer that all users purchase their own TOPS labs, we accept that real-life situations sometimes call for flexibility.

Reselling, trading, or loaning our materials is prohibited unless one or both parties contribute an Honor System Royalty as fair compensation for value received. We suggest the following amounts – let your conscience be your guide.

HONOR SYSTEM ROYALTIES: If making copies from a library, or sharing copies with colleagues, please calculate their value at 50 cents per lesson, or 25 cents for homeschoolers. This contribution may be made at our website or by mail (addresses at the bottom of this page). Any additional tax-deductible contributions to make our ongoing work possible will be accepted gratefully and used well.

Please follow through promptly on your good intentions. Stay legal, and do the right thing.

SCHOOLS, DISTRICTS, and HOMESCHOOL CO-OPS:

PURCHASE Option: Order a book in quantities equal to the number of target classrooms or homes, and receive quantity discounts. If you order 5 books or downloads, for example, then you have unrestricted use of this curriculum for any 5 classrooms or families per year for the life of your institution or co-op.

2-9 copies of any title: 90% of current catalog price + shipping.

10+ copies of any title: 80% of current catalog price + shipping.

ROYALTY/LICENSE Option: Purchase just one book or download *plus* photocopy or printing rights for a designated number of classrooms or families. If you pay for 5 additional Licenses, for example, then you have purchased reproduction rights for an entire book or download edition for any **6** classrooms or families per year for the life of your institution or co-op.

1-9 Licenses: 70% of current catalog price per designated classroom or home.

10+ Licenses: 60% of current catalog price per designated classroom or home.

WORKSHOPS and TEACHER TRAINING PROGRAMS:

We are grateful to all of you who spread the word about TOPS. Please limit copies to only those lessons you will be using, and collect all copyrighted materials afterward. No take-home copies, please. Copies of copies are strictly prohibited.

CONTENTS

GETTING IT TOGETHER

You hold within your hands a **complete teaching resource.** This book contains 20 reproducible hands-on science lessons together with all necessary information to help you teach each lesson successfully. All you add are the simple materials listed at the bottom of the page.

Look it over. This modest list contains everything you need to teach **every** lesson. Most of the materials you already have. Get the rest from your local supermarket or have your students bring the required items from home.

Each item is **listed in order** of first appearance in the student activities. To start getting it together, begin at the top of this list and work down. Gather everything at once, or collect materials as your students progress through each lesson.

Needed quantities depend on several factors: how you teach, how many students you have and how you organize them into activity groups. The 3 numbers listed by each item correspond to the main teaching strategies in use today. Find the one that suits your teaching style and gather quantities accordingly.

From time to time the teaching notes contain suggestions for additional activities called EXTENSIONS. Materials for these optional experiments are not listed here nor under MATERIALS in the teaching notes. Read instead the extension itself to find out what new materials, if any, are required.

Once you collect the needed materials, place them on an equipment table or on open shelves that are accessible to your students. Items of special value may require a locked cabinet or a special check-out box near the teacher's desk.

Many of the materials you use in this module are used in other TOPS modules as well. As you continue with other TOPS modules and build your inventory, you'll find that gathering materials requires less and less effort!

Q₁ — Resource Center / Activity Corner / Parent-Child Activity / Demonstrations

There is enough material so that 1 student or group of students can complete all the activities.

If you multiply Q1 by 2, then there will be enough materials for two groups to work on the same activity or, perhaps, for three or more groups to simultaneously work on different activities.

Q₂ — Individualized Approach

Initial activities require almost as much duplication as the traditional approach. But quantities soon drop off as groups "spread out" within the module, doing different activities at different times.

Students group naturally and informally according to academic or social preferences. Group membership tends to change as slower members fall back into slower groups and faster members move up into faster groups.

Quantities in Q2 assume a total class size of about 30 students working in 10 groups of 3 each. Modify as necessary to fit your own particular requirements.

Q₃ — Traditional Class Lessons

The teacher introduces each lesson to the class as a whole, then everyone does the activity together. Time at the end of the period is reserved for summarizing and reinforcing key concepts.

Quantities in Q3 again assume a class size of about 30 students working in groups of 3. The numbers are sometimes higher than Q2 because greater duplication of materials is needed when everyone works simultaneously on the same worksheet.

MATERIALS

Q₁	/Q₂	/Q₃	
1	/30	/30	sheets of lined notebook paper
1	/20	/30	pairs of scissors
1	/30	/30	wooden spring-action clothespins
1 roll			masking tape
3	/1 box		paper clips
1	/30	/30	soda pop cans
1	/10	10	rolls cellophane tape
1 spool			thread
2	/50	/60	pennies
2	/50	/60	3x5 index cards
1 piece			butcher paper (optional - see teaching notes 13)
4	/150	/150	plastic soda straws
2	/60	/60	long straight pins
1 pkg. ea.			pinto beans, popcorn, lentils, long-grained white rice

Q₁	/Q₂	/Q₃	
1	/20	/30	staples
3	/90	/90	disposable cups - paper or styrofoam - 6 oz. or more
1	/10	/10	sources of water - from a sink or pitcher
1	/4	/10	teaspoons - should approximate standard teaspoon capacity
1 sheet			aluminum foil
1 bottle			plain uncoated aspirin - 5 grain tablets
several			pkgs. candy, nuts or raisins - see teaching notes 20

A

SEQUENCING ACTIVITIES

This logic tree shows how all the worksheets in this module tie together. In general, students begin at the trunk of the tree and work up through the related branches. As the diagram suggests, the way to upper level activities leads up from lower level activities.

At the teacher's discretion, certain activities can be omitted or sequences changed to meet specific class needs. The only activities that *must* be completed in sequence are indicated by leaves that are linked vertically with an *open space* in between. In this case the lower activity is a prerequisite to the upper.

When possible, students should complete the worksheets in numerical sequence, from 1 to 20. If time is short, however, or certain students need to catch up, you can use the logic tree to identify concept-related *horizontal* activities. Some of these might be omitted since they serve only to reinforce learned concepts rather than to introduce new ones.

On the other hand, if students complete all the activities at a certain horizontal concept level, then experience difficulty at the next higher level, you might go back down the logic tree to have students repeat specific key activities for greater reinforcement.

For whatever reason, when you wish to make sequence changes, you'll find this logic tree a valuable reference. Parentheses in the upper right corner of each student worksheet allow you this flexibility: they are left blank so you can pencil in sequence numbers of your own choosing.

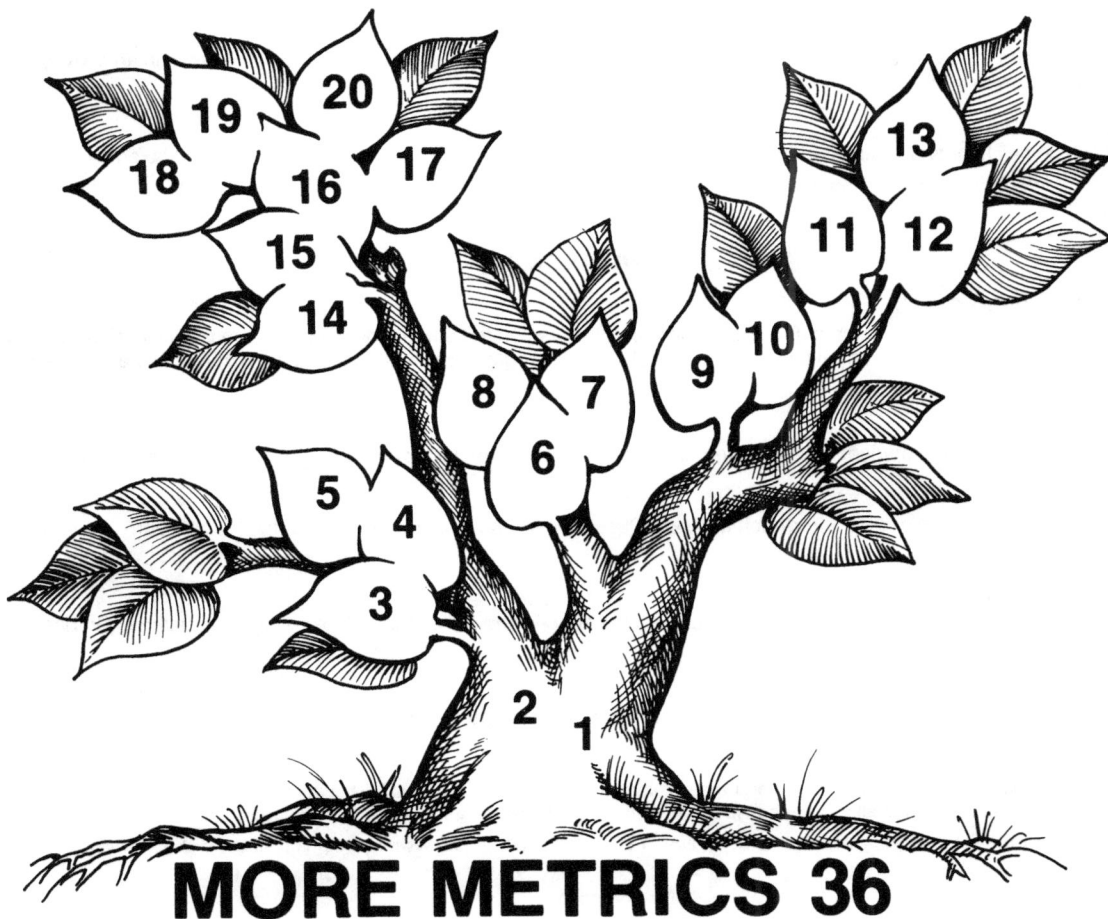

MORE METRICS 36

BUILDING AN EFFECTIVE TEACHING STRATEGY

No teaching strategy is totally effective in all classrooms situations. This module is flexibly arranged to adapt to a wide *range* of teaching possibilities. Design your own strategy: select options listed below that best fit your own needs and meet the needs of your students.

A. Classroom Organization

1 RESOURCE CENTER
Worksheets and science materials are placed in a special resource area. Students come from the classroom to work independently on science activities. Teachers or aides are available to assist students as the need arises.

2 ACTIVITY CORNER
This operates like the resource center, except a special area is designated *within* the classroom itself. Students come here to do science experiments after their regular class work is completed.

3 PARENT-CHILD ACTIVITY
A parent, teacher or aide works with one or more students in a tutorial relationship. This may occur during or after school hours or in the home.

4 TEACHER DEMONSTRATION
The teacher performs experiments in front of the class, inviting occasional student participation. This approach is often used for younger children who do not have sufficient manual dexterity to manipulate the materials.

5 INDIVIDUALIZED ACTIVITY
Students proceed through the worksheets at their own pace. Those working on the same activity informally group together, reducing substantially the total number of experiments going on in the classroom. The teacher acts as a learning supervisor, responding to questions and problems as they arise within the context of class activity. After the most advanced students complete all the worksheets, the class moves on to a new module *together*.

6 TRADITIONAL CLASS TOGETHERNESS
Each activity constitutes a specific lesson to be completed during a specified time frame. The teacher introduces the activity to all the students together, then breaks the class into managable lab groups that each do the same experiment. A class discussion sometimes follows to summarize key concepts and provide lesson closure.

B. Reproduction of Activity Sheets

1 OVERHEAD PROJECTION
Place each worksheet directly on an opaque projector or prepare a transparency.

2 ACTIVITY CARDS
Make 2 or 3 photocopies of each worksheet. Plasticize them to make durable full page activity cards. File these in an activity folder or display them on a bulletin board or wall.

3 WORKSHEETS
Duplicate enough copies to provide each student with a worksheet. These can be photocopied directly or thermofaxed onto a master ditto, then run off on a spirit duplicator. Distribute copies to students directly. Or place each set in a separate folder and file them in a box so students can use them as needed.

C. Evaluation

1 PASS / NO-PASS CHECKPOINTS
Daily write-ups are evaluated by the student and teacher together *in class*. If the student demonstrates reasonable effort commensurate with ability, the write-up is simply checked off, either in a grade book or on a progress chart attached to each student's personal assignment folder kept on file in class.

2 GRADED ASSIGNMENTS
Write-ups are handed in by each student as completed, graded by a teacher or an aide and then returned to the student.

3 QUIZES
The teacher gives a quiz (written or oral) after each activity. Questions for the quiz are taken from the "Evaluation" sections in the teaching notes.

4 INFORMAL OBSERVATION
The teacher takes mental note of active participators who work to capacity and of inactive onlookers who waste time. Grades are awarded accordingly.

5 EXAMS
An exam given to all students at the same time covers key concepts from activities that all students have completed and reviewed. Questions come from the "Evaluation" sections in the teaching notes.

Among the teaching options listed above, we recommend the combination A5 - B3 - C1 - C4 - C5. This approach combines elements from two opposite teaching strategies in a most effective way: it allows for individual differences while maintaining traditional class togetherness.

Would such an approach work in your own classroom? The "Dairy of a Teacher" which follows may help you visualize the answer. It is based on my own classroom experience.

C

DIARY OF A TEACHER

THE DAY BEFORE

Tomorrow is the first day of school. With anxiety and anticipation you check to see that everything in your classroom is in good order.

You have already duplicated 30 copies each of the first few lessons in this TOPS module. They lay on your desk, each in a manila folder marked with large numbers 1, 2, and 3. You make a mental note to bring a large box and a brick to school tomorrow. These will prop the folders upright like a vertical file, and help keep you organized as you add additional folders.

You wonder how you ever managed without manila folders. Already you have printed each student's name on a fresh new assignment folder and stapled a sheet of graph paper to the inside cover to track each student's progress. When they arrive tomorrow, you will surprise them with a worksheet, a file folder and the simple instruction, "Get busy." You've already laid out the necessary materials on a table in the back of your room. You smile inside yourself; you haven't felt quite this prepared in years.

DAY 10

Your class has been humming now for 10 straight days. Perhaps not humming: buzzing more aptly describes the state of orderly confusion. Students have questions and problems to be sure. (You wish they would at least *read* the instructions before running to you for an explanation.) Still, the worksheets provide a firm sense of direction. Students know where they are and understand where they need to go.

Now that students understand your system, they come to class and get straight to work on what they were doing the day before. Just before lunch they tend to quit early, but at other times you have to pry them away from their experiments. You tell the slower ones to assign themselves homework to catch up and it seems to be working!

The assignment folders work well too. Students point with pride at the growing list of check-points you have marked off on their graph paper progress charts. As their folders expand so does their self confidence.

DAY 15

Today 2 groups of students who seem to be racing each other have completed all 20 activities. The bulk of your class remains 3 to 6 activities behind, with a few stragglers plus the new kid bringing up the rear.

You announce that individualized worksheet activity will end in 2 days. The most advanced students seem eager to work on several experiments of their own. You can follow that up with an "Extension" activity if time allows. You help the slow ones catch up by assigning three key-concept activities while skipping the rest.

There is a frenzy of activity as students rush to meet your deadline. They know that part of their grade is determined by the total number of activities they complete.

DAY 18

Today you kick back and relax. You have assigned several students to give reports on their original investigations. The rest of the period will be taken up with a film. For tomorrow you've planned a blackboard review of major module concepts. Then on Friday you'll finish off with an exam.

The kids are already bugging you about grades and asking what they will be studying next. You decide to give them a 3-part grade weighted equally on pace (number of lessons completed), attitude, and the exam. As to what they'll study next, you can't decide. Perhaps you'll let *them* decide.

Its already the fourth week into the school year and you don't even feel the strain. Activity-centered teaching seems so natural and easy. You respond to questions that kids have instead of the other way around. You ask yourself why you never taught this way before.

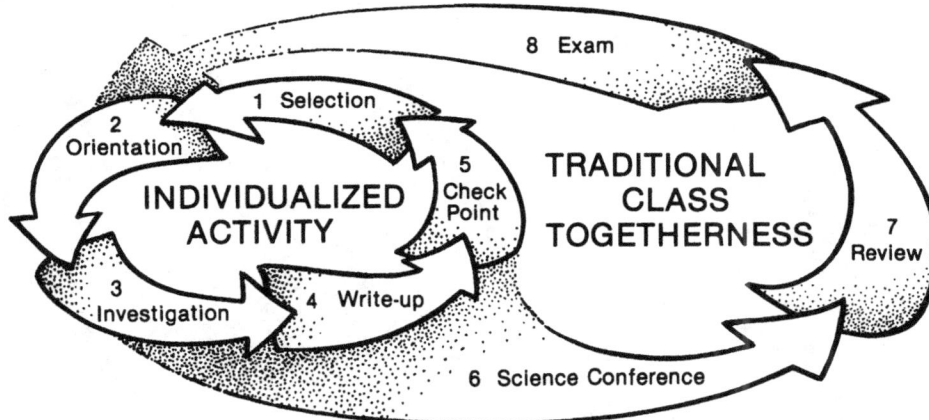

1. **SELECTION.** Students generally select worksheets in the order you specify. They should be allowed to skip a task that is not challenging, however, or repeat a task with doubtful results. When possible, encourage students to do original investigations that go beyond or replace particular activites.

2. **ORIENTATION.** Good students will simply read worksheet instructions and understand what to do. Others will require further verbal interpretation. Identify poor reader in your class. When they ask, "What does this mean?" they may be asking in reality, "Will you please read these instructions aloud?"

3. **INVESTIGATION.** Students observe, hypothesize, predict, test and analyze, often following their own experimental strategies. The teacher provides assistance where needed and the students help each other. When necessary, the teacher may interrupt individual activity to discuss problems or concepts of general class interest.

4. **WRITE-UP.** Worksheets ask students to explain the how and why of things. Answers should be brief and to the point. Students may accelerate their pace by completing these out of class.

5. **CHECK POINT.** The student and teacher evaluate each write-up together on a pass/no-pass basis. If the student has made reasonable effort consistent with individual ability, the write-up is checked off on a progress chart and included in the student's personal assignment folder kept on file in class.

6. **SCIENCE CONFERENCE.** After individualized activity has ended, students come together to discuss experiments of general interest. Those who did original investigations give brief reports. Slower students learn about the later activities completed only by faster students. Newspaper articles are read that relate to the topic of study. The conference is open to speech making, debate, films, celebration, whatever.

7. **REVIEW.** Important concepts are discussed and applied to problem solving in preparation for the module exam.

8. **EXAM.** Evaluation questions are written in the teaching notes that accompany each activity. They determine if students understand key concepts developed in the worksheets. Students who finish the test early begin work on the first activity in the next new module.

LONG-RANGE OBJECTIVES

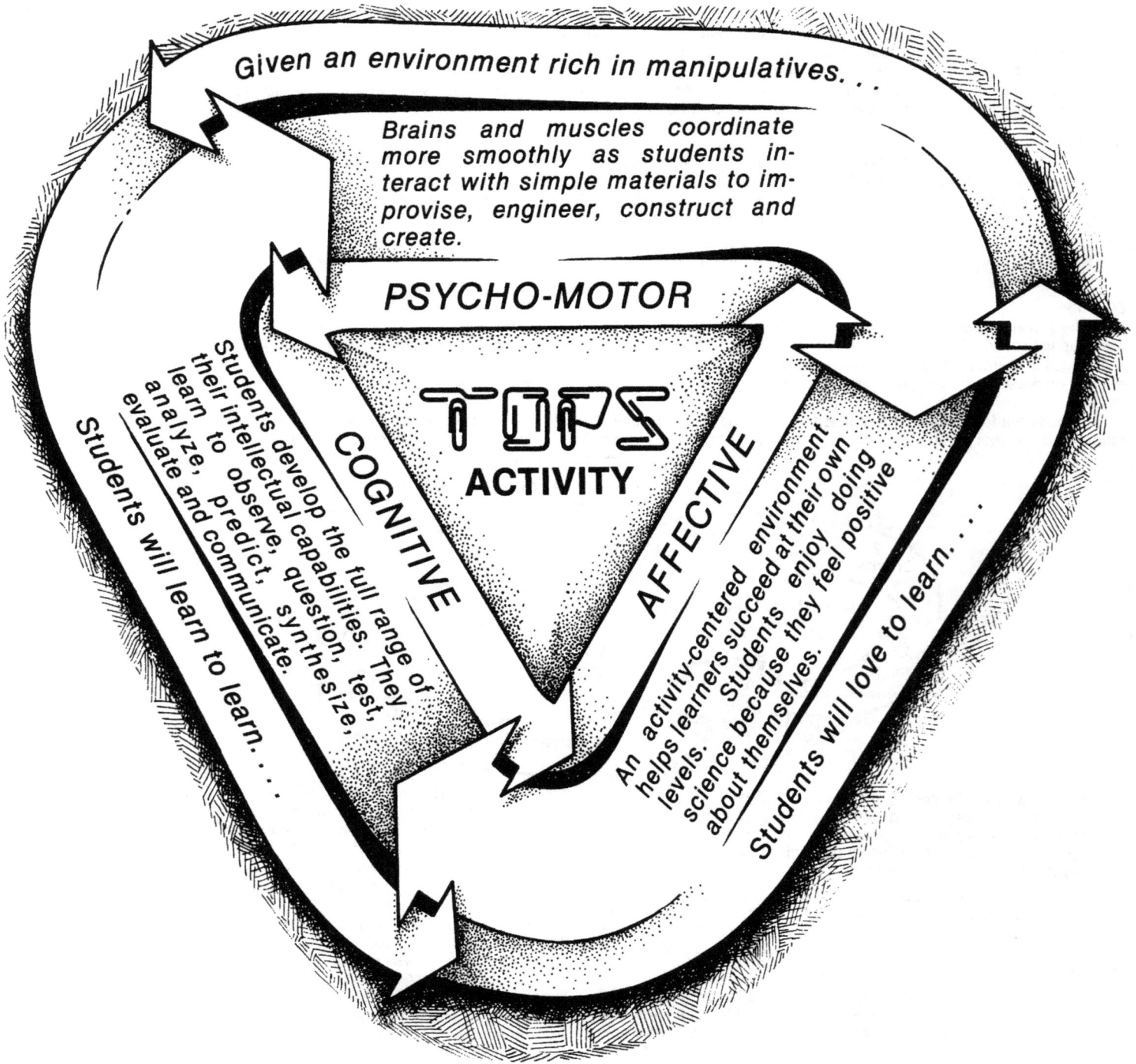

Given an environment rich in manipulatives. . .

Brains and muscles coordinate more smoothly as students interact with simple materials to improvise, engineer, construct and create.

PSYCHO-MOTOR

TOPS ACTIVITY

COGNITIVE

Students develop the full range of their intellectual capabilities. They learn to observe, question, test, analyze, predict, synthesize, evaluate and communicate.

Students will learn to learn. . . .

AFFECTIVE

An activity-centered environment helps learners succeed at their own levels. Students enjoy doing science because they feel positive about themselves.

Students will love to learn. . . .

E

Science is an interconnected fabric of ideas woven into broad and harmonious patterns. Use "Extensions" in the teaching notes plus the outline presented below to help your students grasp the big ideas—to appreciate the fabric of science as unified whole.

Imagine that you are living long, long ago. Nobody in your culture has ever dreamed about measuring anything before. Write a story telling how you **invented the process of measuring for the first time.**

How long is a meter? Who first decided? How is it defined today? Go to the library to **research the history of the meter.**

Related TOPS modules that provide additional hands-on measuring experience using simple materials include:

02 Measuring Length
03 Graphing
06 Metric Measure
35 Metric Measuring

MORE METRICS 36

Have inventors ever been able to improve upon the accuracy of a ruler? Write a report on the **vernier caliper.**

Can you **list 100 measuring units that are all different?** Go for a record! (No fair writing the same unit with a different prefix. If you list "meter" you can't list "centimeter". If you write "inch" you can't write "half inch".)

An **astronomer**, a **surveyor**, and a **pharmacist** are all together at a party "talking shop". What kind of confusion might result? Write a dialogue.

How big is our planet? How far around? What is the capacity of our oceans? **Write a book of Earth Facts.** Include units of measure that are easy to visualize and understand.

TEACHING NOTES
For Activities 1-20

Teaching Notes 1

Write this metric vocabulary, and a dollar sign, on 4 index cards folded to stand upright. Use large bold lettering, visible from the back of your room. Display these cards on a table in no particular order) together with a dollar bill, a dime, a penny and 1/10 penny. (Represent 1/10 penny by covering all but a tenth of it with masking tape.)

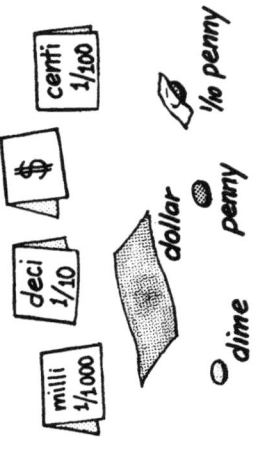

milli 1/1000   deci 1/10 $ centi 1/100

dollar   dime   penny   1/10 penny

Match the dollar bill with the dollar sign. Then challenge your class to pair the other 3 coins with their correct prefix classification.

$	deci	centi	milli
dollar	dime	penny	1/10 penny

Next, place all money and signs in descending order. Holding up any one, ask your class to name the other 3 in terms of the one in your hand.

Repeat for all 4 denominations.

Evaluation

Q. If this line measures a "centi".

a. Draw a "milli" here: _____
b. Draw a "deci" here: _____

A: a. (Students should draw a line ten times *shorter* than the "centi" line above.)
b. (Students should draw a line ten times *longer* than the "centi" line above.)

Materials

☐ Lined notebook paper.
☐ Scissors.
☐ Clear tape.
☐ A clothespin.
☐ A sturdy table or chair. Your students will need to reach the ceiling safely in step 6.

More Metrics 36 is all new — totally different from **Metric Measuring 35**. It provides an additional 20 lessons that will help your students understand this all-important measuring system. Like all modules in our **Science With Simple Things** series, **More Metrics 36** has no prerequisites. We always start with basic principles and build from there. You can teach this module independently or as a sequel to **Metric Measuring 35**.

4b. To make 10 layers, it's much easier to fold 1 strip into 10 parts than to cut each piece individually.

4 layers
6x16 layers
100 layers

Less coordinated students may need help. These many layers tend to "squirt" out of the clothespin jaws unless placed firmly inside.

5. Notebook paper is generally quite thin. We estimate that 1000 sheets will stack about 8 cm high. Your students may draw lines longer than this because they fail to compensate for the folds in the paper that flare out on each side.

MORE THAN A DECI

A TRUE DECI

6. Allow each student to follow an independent strategy. Some may cut paper to size; others will want to use string or thread. Mathematicians might measure the wall, then divide by the length of the line.

7. To solve this problem, it is essential that students understand decimal multiplication and division. Write a row of zeros on your blackboard with a "1" in the middle. Use a pencil to place your decimal point. Ask your students to count in unison as you move the pencil right or left. Your class should chorus "one, ten, hundred, thousand . . ." as you multiply right and, "one, tenth, hundredth, thousandth . . ." as you divide left.

0000001.0000000
MILLIONTH   ONE   TEN THOUSAND

Discussion

If milli, centi and deci are unfamiliar terms, relate them to something that is familiar — money.

Task Objective (TO) understand metric prefixes as simple multiples of ten. To estimate numbers by comparing lengths.

MILLI, CENTI, DECI & MORE

1 Cut off 7 strips of notebook paper along the blue lines.

ONE SPACE WIDE

2 Fold one strip in half 4 times, then unfold it. This makes 16 equal parts.

16 parts count them!

3 Cut off just 1/16 and . . .

tape it here
milli

LABEL YOURS "MILLI"

Let's call this a MILLI.

4 One *milli* is only one paper thick. Open a clothespin clip . . . Now open your clip.

— 1 MILLI wide. —
1 paper layer

Ten *milli* make one *centi* . . . .

— 1 CENTI wide. —
10 paper layers

Ten *centi* make one *deci* . . . .

— 1 DECI wide. —
100 paper layers

USE YOUR DECI TO ESTIMATE.

5 Draw a line showing how far 1000 layers will reach.

(approximately 8 cm)

6 Estimate how many sheets of paper you can stack from floor to ceiling. Write how you did it.

Cut a strip of paper to size. Use it to count up the wall 1000 sheets at a time.

A ceiling 2½ m high contains roughly 30,000 sheets.

7 *Stack to the moon?*

If 1,000 papers stack about 8 cm, how many reach 4 cm?

ABOUT 400,000 KM

(Half the distance)   **500 papers**

How many sheets reach 4 meters?
(100 times farther)   **50,000 papers**

How many sheets reach 4 kilometers?
(1,000 times farther)   **50,000,000 papers**

How many sheets of paper would stack to the moon?
(100,000 times farther) . . . **5,000,000,000,000 papers**
(5 trillion) . . .

TOPS LEARNING SYSTEMS

2. When students complete this worksheet, they should come to you, as usual, for a check point. Don't expect too many to score a perfect 36 out of 36. But you can turn the ones missed into a positive learning experience by following this simple strategy.

While each student looks on, circle each wrong answer on the worksheet. Ask the student, in turn, to circle where the right answer can be found on the ruler directly above. Knowing the position of each answer, it's relatively easy to count the correct number of millimeters, centimeters or meters back to zero.

Expect many perfect scores the second time around. Perhaps a triple-check or quadruple-check will be necessary for some. Don't dwell on how many check points it takes to get all the answers right. Celebrate, instead, the final perfection!

3. These 7 relationships are worth committing to memory. They are as fundamental to the metric system as multiplication facts are to arithmetic.

Discussion

This ruler stretches 1 full kilometer across your paper and imagination. All 7 units are clearly shown along the way. In a single glance you can appreciate the logic and structure of the whole metric system.

Such a remarkable ruler deserves special study. Consider leading your class on a pencil-point tour. As tour guide, you call out positions on your ruler. Then ask your class to point to each position you call, on their own worksheet rulers, with their pencil points.

Begin by reviewing the smallest decimal divisions: 1 mm, 2 mm, 3 mm... Circulate about your room to see that everyone is keeping up...8 mm, 9 mm, 10 mm... Past 1 cm everyone must visualize the mm divisions. This is good practice for later exercises in estimating. Count by larger and larger mm intervals... 15 mm, 24 mm, 100 mm, 230 mm, 1,000 mm, 10,000 mm, 100,000 mm... Finally call out 1 million mm, reaching the kilometer crest on the distant hill.

Next take a tour of centimeters: 1 cm, 2 cm, 4.5 cm...80 cm, 100 cm, 200 cm, 300 cm, 1,000 cm, 10,000 cm, 100,000 cm. Then tour meters: .01 m, .02 m, .03 m, .1 m,....9 m, 1 m, 10 m, 100 m, 1,000 m. Avoid measures that are not well represented in the drawing. Four meters, for example, hides behind the first hill.

As you walk through the room, you'll easily recognize those who are pointing their pencils to the correct places, and those who need special help.

Even gram and liter relationships can be inferred from this same graphic. If 1,000 mm make a meter, then in like manner 1,000 ml make a liter and 1,000 mg make a gram. Students who forget metric equivalents in later activities should return here as many times as necessary to refresh their memories.

Materials

None.

Evaluation

Q: How many millimeters in ....

1 cm .... ?
6 cm .... ?
7.5 cm .... ?

How many centimeters in ....

1 dm .... ?
20 mm .... ?
2 mm .... ?

A:
1 cm = 10 mm
6 cm = 60 mm
7.5 cm = 75 mm

1 dm = 10 cm
20 mm = 2 cm
2 mm = .2 cm

(TO) visualize how metric units fit together as multiples of 10. To practice expressing one measure in terms of another.

NAME:

CLASS:

More Metrics ( )2

KILOMETER RULER

1 Find 7 units of measure on this ruler. Write them in order here:

LONGER
kilometer
hectometer
dekameter
meter
decimeter
centimeter
millimeter
SHORTER

each unit is ten times...

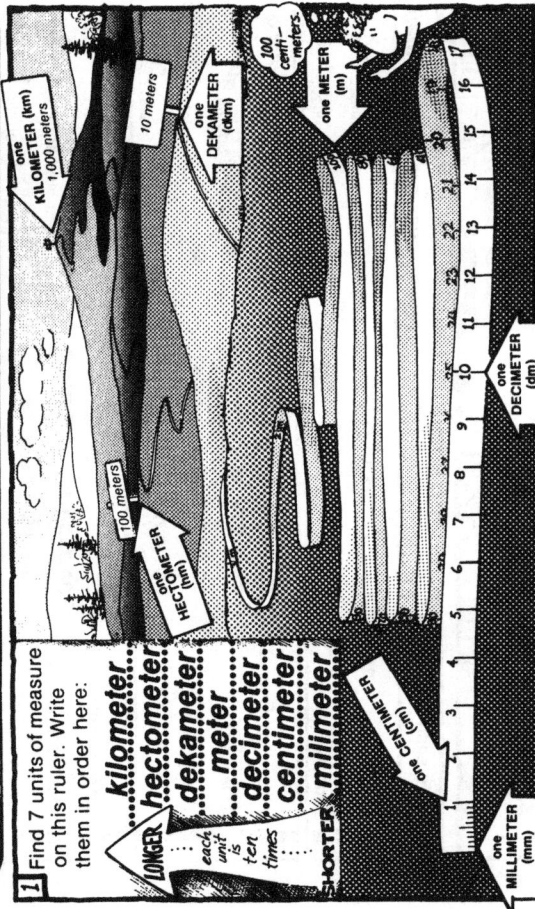

one KILOMETER (km) 1,000 meters
one HECTOMETER (hm) 100 meters
10 meters
one DEKAMETER (dkm)
one METER (m)
100 centimeters
one DECIMETER (dm)
one CENTIMETER (cm)
one MILLIMETER (mm)

3 List the 7 stars here:

1 cm = 10 mm
1 m = 1,000 mm
1 mm = .1 cm
1 m = 100 cm
1 cm = .01 m
1 mm = .001 m
1 km = 1,000 m

These are used often! Memorize then!

2

How many millimeters in ..		How many centimeters in ..		How many meters in ..	
1 cm?	10 mm	50 mm?	5 cm	100 cm?	1 m
5 cm?	50 mm	35 mm?	3.5 cm	50 cm?	.5 m
2.5 cm?	25 mm	5 mm?	.5 cm	20 cm?	.2 m
7.8 cm?	78 mm	1 mm?	.1 cm	19 cm?	.19 m
70 cm?	700 mm	250 mm?	25 cm	10 cm?	.1 m
1 dm?	100 mm	1 dm?	10 cm	3 cm?	.03 m
3 dm?	300 mm	9 dm?	90 cm	1 cm?	.01 m
1.6 dm?	160 mm	.5 dm?	5 cm	5 mm?	.005 m
1 m?	1,000 mm	1.2 dm?	12 cm	1 mm?	.001 m
4 m?	4,000 mm	1 m?	100 cm	300 mm?	.3 m
1 dkm?	10,000 mm	1 dkm?	1,000 cm	1 km?	1,000 m
1 km?	1,000,000 mm	1 hm?	10,000 cm	5 km?	5,000 m

Teaching Notes 3

Enlarge all 28 gray and black squares to serve as flash cards. They ask 28 important questions. Print each one in large bold letters on a quarter sheet of standard-size paper. The white squares supply 14 possible answers. Write the correct response on the back of each flash card to serve as a quick answer-key reference.

SAMPLE FLASH CARDS:

ONE millimeter

As thin as a dime

ONE kilogram

1000 g

2. These TOPS peoplets are acting out 3 distinct forms of measure. If your students are metric beginners, they may benefit from doing a similar kinesthetic exercise. Call out any one of 3 kinds of measure:

volume   mass   length
liter (or) gram (or) meter
milliliter   kilogram   centimeter

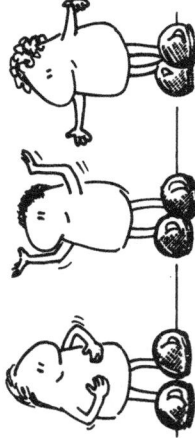

Ask your class to motion back the correct response. Continue faster and faster until it all ends in buckets of laughter.

This step must be completed accurately. Otherwise each pile will not properly subsort into equal triplets in step 3. The volume pile contains just 6 squares with 2 grays reading, "As *much as* ...." The mass pile contains 12 squares with 4 grays reading, "As *heavy* as ...."

3-4. Step 3 requires plenty of free desk space or an open floor. Sort the volume squares first. This is easy to do because there are only 2 triplets. Next sort the 12 mass squares into 4 triplets. Finally the 24 length squares into 8 triplets. It's easiest to lay down all the white squares first. Then pair up the grays, and finally match the blacks.

Record your results in step 4 as you go, or after you've matched all 14 triplets.

Remind your students to hang onto their worksheets as well as their squares. The worksheet provides a handy reference should students forget some of the metric relationships. The squares serve as game cards in later activities.

Evaluation

Q: Write the correct letter in each blank to make a match.

a. gram
b. milliliter
c. kilogram
d. millimeter
e. centimeter
f. liter
f. milligram
h. kilometer
i. meter

    As much as 3 cans of pop
    As far as a 10 minute walk
    As thin as a dime
    As much as 1/5 teaspoon
    As heavy as a textbook
    As thick as a slice of bread
    As long as a pace
    As heavy as 2 paper clips
    As heavy as a fruit fly

A:

(liter)	f	As much as 3 cans of pop
(kilometer)	h	As far as a 10 minute walk
(millimeter)	d	As thin as a dime
(milliliter)	b	As much as 1/5 teaspoon
(kilogram)	c	As heavy as a textbook
(centimeter)	e	As thick as a slice of bread
(meter)	i	As long as a pace
(gram)	a	As heavy as 2 paper clips
(milligram)	g	As heavy as a fruit fly

Drill your students with these flash cards often — as an introduction to this activity, as a review when you finish, and as periodic reinforcement throughout this module.

Discussion

You want your class to get comfortable with the metric system — to instantly recognize important units as familiar old friends. Students need to relate to meters, liters and grams as easily as they relate to feet, quarts and pounds. If they don't, they'll stop using metrics just as soon as you stop teaching this module. That would be a shame.

Metric Squares contain most of the necessary facts your students need to know. You can help them assimilate this information with an old-fashioned flash card drill.

Materials

☐ A reproduced sheet of metric squares.
☐ Scissors.

NAME: _____   CLASS: _____

More Metrics ( )3

(TO) become familiar with 42 important interrelated facts about metric volume, mass and length.

METRIC SQUARES

1 Get a sheet of METRIC SQUARES. Cut out all 42 squares *and* 3 labels.

2 Sort your squares into 3 labeled piles:

VOLUME   6 SQUARES
MASS   12 SQUARES
LENGTH   24 SQUARES

3 Now sort each pile into equal triplets—groups of 1 white, 1 grey, and 1 black that are equal.

VOLUME (2 triplets)
MASS (4 triplets)
LENGTH (8 triplets)

TRIPLETS are always EQUAL

Record each set below!

4 Write each triplet in the correct space below.   **(some answers interchangable)**

VOLUME
One liter is ...... as much as 3 cans of pop ...... and equals.. 1000 ml
One milliliter is ...... as much as 1/5 teaspoon ...... and equals.. .001 *l*

MASS
One kilogram is ...... as heavy as a textbook ...... and equals.. 1000 g
One gram is ...... as heavy as 2 paper clips ...... and equals.. .001 kg
One gram is ...... as heavy as 2 raisins ...... and equals.. 1000 mg
One milligram is ...... as heavy as a fruit fly ...... and equals.. .001 g

LENGTH
One kilometer is ...... as far as a 10 minute walk ...... and equals.. 1000 m
One meter is ...... as wide as a doorway ...... and equals.. 1000 mm
One meter is ...... as long as a pace ...... and equals.. 100 cm
One meter is ...... as long as a monkey's tail ...... and equals.. .001 km
One centimeter is ...... as thin as a slice of bread ...... and equals.. 10 mm
One centimeter is ...... as wide as a fingernail ...... and equals.. .01 m
One millimeter is ...... as thin as a dime ...... and equals.. .1 cm
One millimeter is ...... as long as a flea ...... and equals.. .001 m

SAVE your metric squares!

(TO) firmly link metric units with common conversion factors and concrete images.

NAME: _____
CLASS: _____

More Metrics ( )4

FACE-UP

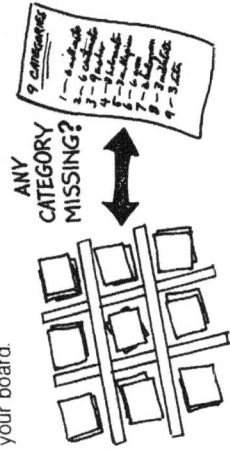

1 Lay out a grid of masking tape on your desk so your metric squares can fit inside.
Like tic-tac-toe

2 Shuffle your deck of 42 cards to mix them well.
Then — 2 equal cards?

3 Place the top 10 cards *face up* on your grid.
Start with 2 cards in the middle...
...1 card everywhere else.

4 Search for 2 equal cards among the 9 on your grid.
DRAW PILE (face up)
... Cover each pair you find with 2 new cards from your draw pile. Put them face up so they can form new pairs.
Equal!

5 Keep playing until ... you win! (all cards played).
DRAW PILE GONE
Matched them all!
... you lose! (cards left in draw pile)
Stuck...
I see a pair!

6 Now try these games:

SOLITARY FACE-UP
Repeat steps 2-5 on your own.
TEACHER ✔: ☐☐☐

COOPERATIVE FACE-UP
1 mg is as heavy as a flea.
Take turns finding 2 equal cards. Say each pair out loud.
your turn...
TEACHER ✔: ☐☐☐

COMPETITIVE FACE-UP
100 cm = 1 meter!
RATS — now we're TIED!
Try to be first to call out pairs. Keep score.
TEACHER ✔: ☐☐☐
SAVE your squares

TOPS LEARNING SYSTEMS

In practice, beginners make many errors. They don't see pairs that really are there. They do see pairs that aren't — an easy way to get unstuck. Whether they win or lose is irrelevant. It's how they play the game. Those who learn are playing just as they should.

6. Some students thrive on the excitement of competition; others abhor it. Some work cooperatively together; others work best alone. In this step you can have it your way: play all 3 games or concentrate just on the ones you like. Remind your class to stay on task until they collect the number of game-completion teacher-checks that you assign. We suggest that you require three.

Notice that both Cooperative and Competitive Face-Up ask students to call out each discovered pair aloud. This is important for several reasons. First, learning is enhanced with audio feedback. Second, students may recognize numbers without being able to read them aloud. They need practice saying "one thousandth of a gram" when they see .001 g. Third, calling each pair out loud allows an opponent to double check another's answer, resulting in fewer game errors.

Have metric recognition skills in your class improved as result of playing Face-Up? You can assess progress by drilling once again with the flash cards you made in the last activity. Recognition times by now should be shrinking below 1 second. Instantaneous recognition is your goal.

1. Lay out this table grid with masking tape, not clear tape. Masking tape is easy to see, and easy to peel off the table when you finish.

3. There are 42 cards in the deck. Placing an extra in the middle leaves an even 32 in your hand, instead of an odd 33. This means you can discard pairs until you're left with none, instead of 1.

4. The object is to match equal pairs. Some students may be thinking in terms of triplets, remembering the last activity.

5. There are 9 categories of metric cards. Each category contains from 3 to 9 matching cards. Here's the breakdown:

6	millimeter
6	centimeter
9	meter
3	kilometer
3	milligram
6	gram
3	kilogram
3	milliliter
3	liter
42	metric squares.

If you can recognize pairs, your chances of winning are excellent. Generally 1 to 3 pairs are available for matching at all times. However, you can occasionally lose. This happens when all 9 categories turn up — a different one in each of the 9 tic-tac-toe positions. In our experience, this kind of no-match board occurs in about 1 out of 5 games. Perhaps someone can calculate for us the precise theoretical probability.

It's not easy to find a *genuine* no-match board. To confirm that no pairs exist, you have to mentally eliminate 36 different pairing combinations. Usually a pair is hiding somewhere on the board that you simply overlooked.

A true no-match board has 1 (and only 1) card from each of the 9 categories. You can easily verify that there are no pairs by taking this simple inventory: check off each tic-tac-toe position against your list of 9. If just 1 category is missing, then there must a match hiding somewhere on your board.

Evaluation

Q: Fill in each blank with the most logical metric unit.

a. Maria lives 4 ....from school.
b. Kim weighs 50 ....
c. Mohammad is almost 2 ....tall.
d. A watermelon seed weighs about 60 ...
e. Jason's hand measures 13 .... long.
f. Gloria had to swallow 5 ..... of medicine all at once.
g. A nickel weighs about 5 .....
h. The crack in the sidewalk is 5 .... wide.
i. Jordan's car holds 40 ....of gas.

A: a. km   d. mg   g. g
        b. kg   e. cm   h. mm
        c. m   f. ml   i. l

ANY CATEGORY MISSING?

Materials

☐ Masking tape.
☐ A set of metric squares.
☐ A paper clip.

in your hand, it may be possible to play all your cards at once. Then you can rummy while your surprised opponent still has many cards. However, should your opponent go out first, the cards you hold count against you. It's dangerous to hold your triplets too long.

As you lay a triplet on the table, be sure to read each card aloud. This allows your opponent to verify an accurate match. This is important. Any mismatched triplet, played in error now, creates other cards in the deck that have no match at all. Double-checking helps insure that the game will play out to an eventual rummy.

5. DISCARD. Traditionally, your last card must be discarded to rummy; it can't be played as a matched triplet. Beginners will play the last card any way they can, according to Hoyle or not. You can introduce this finer rule later if you wish or decide to ignore it completely.

In real rummy you play a match of several games until the winner reaches some arbitrary cumulative total. Here, we've opted for a shorter version: the highest score wins. Try this short version first. If enthusiasm remains high, you can switch to an arbitrary-total scoring method. Simply agree on a winning score before you start. A total of 30 requires perhaps 2 or 3 games to reach.

Extension

Hold a metric tournament. Award the winner a metric ruler!

Evaluation

Q: Draw a circle around each meter.
Draw a square around each cm.
Draw a triangle around each mm.

A:

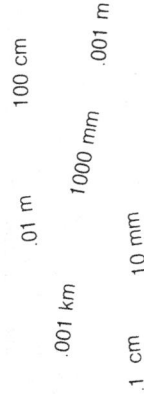
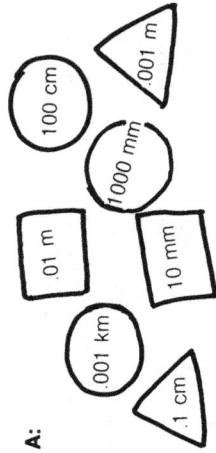

.001 km   1 m   1000 mm   100 cm   .001 m
.01 m   10 mm   .1 cm

Materials

☐ A reproduced metric card holder.
☐ Scissors.
☐ A set of metric squares.

When was the last time you learned to play a card game by reading a book of Hoyle? Chances are, you learned most of the games you know by looking over someone else's shoulder. While it is possible for your class to learn Metric Rummy by studying this worksheet (we think the rules are very clear), a demonstration game will help your students learn much faster.

2. Those who fold the first flap up instead of down are starting off on the wrong foot. This will throw the whole fold sequence out of step, reversing all the peaks and valleys.

3. Any number of students can play Metric Rummy. So players can take turns more frequently, we recommend 2 players per game, or 3 at most.

4. Each card fits into 1 of 9 categories on the card holder. Once classified in the correct position, you can identify triplets at a glance. Unfortunately, your opponent can infer what cards are in your hand as well, by astutely observing how they are positioned. Advanced players can disguise their hands by intentionally placing certain cards in the wrong category.

5. DRAW. Beginners tend to overlook potential triplets in the discard row. So it grows longer than it should. Stress how easy it is to capture these old cards. They are all fair game. Just form an instant triplet from the cards you already hold or the ones among your captured discards.

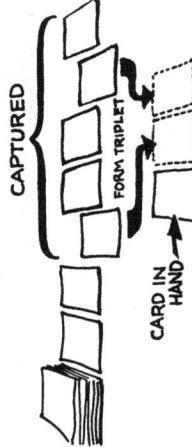

CAPTURED — CARD IN HAND — FORM TRIPLET

It is only fair to first show your opponent how you'll form the triplet, *before* you scoop up your captured cards.

Some games last long enough to empty the draw pile. If this happens, continue play without it. Unless you reach back to match an instant triplet, this means you'll always have to draw the last discard. As long as all players in the game don't draw, then discard the *same* card, the cards should continue to circulate and lead to an eventual winner. If a discard stalemate does occur, enforce this special no-draw-pile rule: the card you draw can't be discarded again until your next turn.

5. MATCH. Except for playing an instant triplet to capture old discards, you don't have to play your other triplets unless you want to. By holding them

---

(TO) memorize metric relationships in a fun way.

METRIC RUMMY

1 Cut a METRIC CARD HOLDER around the outside dashed line.

2 Fold it down, then up, like a fan. The words VOLUME, MASS and LENGTH all fold *down*.
(PEAKS, VALLEYS, FOLD DOWN, METRIC CARD HOLDER, "This is my 'hand.'")

3 Find a friend with a card holder and a deck of 42 shuffled cards.
("Let's play metric rummy!" "OK." 42 SHUFFLED CARDS)

4 Deal 7 cards each, face down, and put them in your card holders ... put the rest, face down, in a draw pile. Turn the top card up to start a discard row beside it.
(DRAW PILE — DISCARD ROW)

5 Take turns. Always follow 3 steps for each turn.

1 draw — Draw 1 card from the draw pile or the discard row. Take only the *newest* discard ... unless an older one forms an *instant triplet* in your hand.
(OLDER, NEWEST, COMPLETES A TRIPLET, TAKE ALL THESE, TOO)

2 match — Match all the equal triplets you can. Lay these in front of you, reading each card aloud.
(WHITE, GRAY, BLACK) "A meter ... As long as a monkey's tail ... 100 cm!" "Those are right..."

3 discard — Put one of your remaining cards at the *end* of the discard row. If you discard your *last card*, shout "rummy" and score!
("RUMMY!!" LAST CARD)

SCORING

● Add 1 point for each card you played in a triplet.
● Subtract 1 pt. for each card left in your hand.
● *HIGHEST SCORE WINS!*

(TO) learn how to read a ruler accurately, estimating the last digit.

NAME:
CLASS:

More Metrics ( )₆

THE LAST DIGIT

1 Write the correct measure in each box. Always make the last digit a *zero* when the arrow points directly to a line.

example: **30.60 cm**

ZERO MEANS ON THE LINE.

☐ TEACHER CHECK

a. 36.20 cm	b. 37.50 cm	c. 33.40 cm	d. 40.10 cm	
e. 38.30 cm	f. 42.70 cm	g. 43.80 cm	h. 32.90 cm	i. 34.60 cm · j. 41.00 cm

2 These arrows point *between* lines. Imagine that each space is divided into ten parts, then *estimate* which tenth comes closest to the arrow.

example: **60.3 cm**

I estimate about 3/10 of the way between.

☐ TEACHER CHECK

a. 62.5 cm	b. 70.2 cm	c. 67.9 cm	d. 69.1 cm	
e. 72.4 cm	f. 66.6 cm	g. 63.7 cm	h. 73.3 cm	i. 64.8 cm · j. 71.0 cm

3 Write each of these measures in 4 digits. Be careful about the last digit.

ESTIMATE if between the lines. Write the ZERO if on the line.

☐ TEACHER CHECK

a. 53.40 cm	b. 51.55 cm	c. 42.30 cm	d. 45.90 cm	
e. 47.87 cm	f. 44.77 cm	g. 41.22 cm	h. 43.59 cm	i. 48.95 cm · j. 50.00 cm

TOPS LEARNING SYSTEMS

---

1. Inches are usually subdivided into halves, quarters, eighths and sixteenths. It's OK to measure this way if you don't mind juggling fractions in your head.

Metric units, by contrast, are always divided by ten, the same base as our number system. Reading a centimeter ruler is as easy as counting. To illustrate, read up the decimal scale while your students follow along with their pencil points; 30.0, 30.1, 30.2...Skip forward at random intervals until you cross the entire ruler: 33.6, 35.9, 38.2.....43.9, 44.0. Real easy, huh?

Now let your students practice on their own. The extra zero in each answer may seem useless and strange at first. (Why add on a zero if it has no value?!) The zero is actually an estimated figure. All accurate measurements have one. It means that as near as you can tell, each arrow hits the line.

To keep students on track, check their progress after each step. Extra check-point boxes are provided for this purpose. Notice that all answers in this step have 4 digits. And units are written with each answer.

Be strict about units. Forgetting to write them is a careless habit that students easily fall into. You'll have work hard to break it. One strategy is to make a firm rule: in this worksheet and all others, a unitless measurement is no measurement at all. Simply refuse to accept unfinished worksheets until they are properly completed — numbers *and* units.

When you see numbers written without units refer to them as so many "apples" or "oranges" or something ridiculous. A number by itself is unspecified. It can refer to anything. So call it anything.

ACCURATE MEASURE = ALL FIGURES YOU KNOW for CERTAIN + estimated, therefore UNCERTAIN + more

2. To measure accurately, write down all the figures you know for sure. Then estimate between the lines to get the last digit.

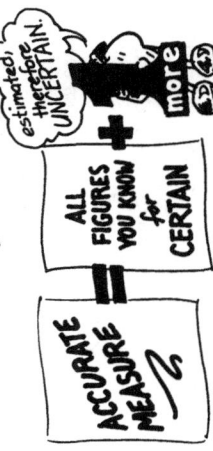

That's all there is to it. Even beginners can measure accurately. You don't have to wait until senior physics!

This time when you hand out check points, be sure that each answer has 3 digits. Allow a variation of ±1 cm in the last digit. After all, it is estimated. And don't forget to check for units. Any proper measure says "how much" and "what".

---

3. Arrows that point to a line end in zero. Arrows that point between lines end in some other number. In either case, the last digit is always estimated.

Consider box j. (It was tricky in the last two steps. This time is no exception.) The arrow points to the mm mark that's sandwiched between 49.9 and 50.1. That must be 50.0. So 50.0 has three certain figures. Does the arrow hit right on the line (50.00)? Does it pull left (49.99)? Does it fade right (50.01)? Because the hundredth place is estimated, you'll never know for sure. Pending more sophisticated technology — a magnifying glass will do — one answer is as worthy as the next.

Evaluation

Q: Fill in each blank with the correct measure.

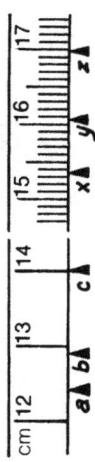

cm |12 |13 |14 |15 |16 |17

a. _____
b. _____
c. _____
x. _____
y. _____
z. _____

A: a. 12.4 cm
b. 12.9 cm
c. 14.0 cm
x. 15.35 cm
y. 16.07 cm
z. 16.98 cm

Metric Clock!

Materials
None.

Teaching Notes 6

1. To estimate between the lines on ruler A, you need to mentally divide the centimeter space into 10 parts, then decide which part comes closest to the arrow. It's not easy to decide because we carefully positioned each arrow to fall **between** imaginary tenths.

Students will experience real uncertainty as they write each answer. It could be the higher; it might be the lower. This is the nature of estimating. You can't tell for sure. However, because the estimated interval is a full centimeter wide, you can say with relative certainty that the arrow lies somewhere between the high and low measure. So student answers should fall in the same range as the key. Allow no extra variation.

Ruler B is different. Here the spaces between each line are narrowed to mm intervals. This increases the visual uncertainty dramatically. Your class may not always estimate the same answers that are one the key. Allow a variation of .01 cm, higher or lower than the answer ranges in the key.

2. The last estimated digit in each measure is always uncertain. If this uncertainty occurs between a nine and a zero, then other digits in the number are also affected. Consider the last box in step 1. Here all the numbers are uncertain because the arrow falls just a whisker to the left of the major 60 cm division. This doesn't imply that the uncertainty is any greater. It's still limited to the hundredths place. Rather, its proximity to a major division creates a kind of ripple effect across the entire number.

Evaluation

Q: Two students measure the length of the *same* index card. One finds its length to be 17.68 cm while the other gets 17.69 cm. Explain why these students get different answers.

A: Each student has measured the index card accurately. All certain figures agree. The answers don't agree in the last digit because it is estimated and thus uncertain.

Materials

None.

--- BALLOT MEASURE ---

BALLOT MEASURE (1)

1. Tape a copy of this arrow to a blackboard or wall.

2. Take a secret ballot: ask your class to, "Write the arrow's position to the nearest cm." Students should estimate from their seats. Don't allow them to approach the figure or take special measurements.

3. Collect all ballots in a hat or box.

4. List the results on your blackboard. You'll get numbers like these:

26 cm	26 cm
26 cm	27 cm
20.6 cm	26.5 cm
26 cm	26 cm

Decide if some measurements were made in error. In our list above we've crossed out three: 20.6 cm is too far to the left of the arrow, a decimal error. 26.5 is wonderfully accurate, but not written to

---

(TO) distinguish between certain figures and uncertain figures. To appreciate that no measurement is exact.

NAME:

CLASS:

More Metrics ( )7

CERTAIN & UNCERTAIN

1 Write 2 possible measures for each arrow.

The # is estimated. It could be .5!

RULER A

example: ↑a.

a. 50.4 cm / 50.5 cm

b. 55.6 cm / 55.7 cm

c. 61.1 cm / 61.2 cm

d. 52.8 cm / 52.9 cm

e. 59.4 cm / 59.3 cm

f. 57.9 cm / 58.0 cm

51.1 is certain, but the last digit is uncertain.

RULER B

example: ↑g.

g. 51.17 cm / 51.16 cm

h. 54.33 cm / 54.34 cm

i. 57.61 cm / 57.62 cm

j. 52.55 cm / 52.56 cm

k. 55.88 cm / 55.87 cm

l. 59.99 cm / 60.00 cm

2 Good measurements have a certain part and an uncertain part. In step 1 above, underline the certain numbers, then circle what's uncertain.

50.4 cm

UNCERTAIN— estimated!

CERTAIN— for sure!

3 Which ruler is more accurate, A or B? Why?

Ruler B is more accurate because it is subdivided into smaller mm divisions.

Can you make a ruler that has no uncertainty? Explain:

No. Even if you subdivide with very tiny, tiny divisions, there is always uncertainty in the spaces between.

4 These folks are having an argument.

It's 51.7! *No! 51.8!*

Can both be right? Explain.

Yes. They only disagree over the last digit. Because it is uncertain, neither one can say for sure.

the nearest cm as directed; 28 cm is too far off the mark to be defended as a good estimate. After eliminating the bad ballots, you'll end up with many 26's and 27's.

5. As you discuss your results ask questions like these:
- Which figure is certain? (2)
- Which figure is uncertain or estimated? (6 or 7)
- Which answer is correct? (Both. The uncertain figures disagree within acceptable limits.)

BALLOT MEASURE (2)

Boldly sketch in the cm divisions with a metric ruler. As you do this, your class can now observe, with certainty, that the arrow lies somewhere between 26 and 27 cm.

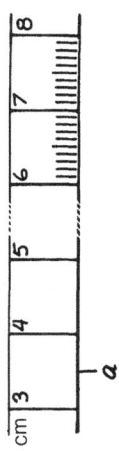
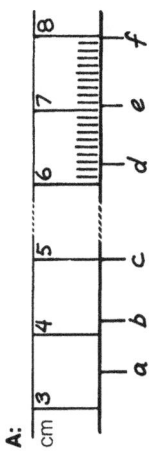

20 cm   **30 cm**

Repeat the balloting procedure again. This time ask students to, "Write the arrow's position to the nearest *tenth* cm."

Discuss your results as before. Ask these questions in addition:
- How has the certainty changed? (There are now 2 certain figures in each answer instead of 1. The uncertainty has shifted from whole cms to mms. Because mms are small, the estimates no longer agree as closely.)
- How can you determine this arrow's position even more accurately? (Pencil in the mm divisions around the arrow, then estimate between these.)
- Is there a limit to how accurately you can know the arrow's position? (Yes. To make the ruler more accurate, you have to keep adding subdivisions. You reach a point where you can no longer see between them.)

6. Work across the table, not down. This allows you to check the accuracy of each pencil mark before drawing the next. This immediate feedback enables you to fine-tune your estimating skills.

Evaluation

Q: Show where each measurement hits the ruler. Draw a thin pencil mark and letter it as shown in the example.

a. 3.5 cm   c. 5.0 cm   e. 7.06 cm
b. 4.2 cm   d. 6.28 cm   f. 7.98 cm

A:

4-5. Here you determine how accurately you marked the ruler in step 2. A proper evaluation requires careful attention to two important details. First, center the mm scale along the cm interval so the ends sit directly over 14 cm and 15 cm respectively. Second, estimate between the mm divisions while holding the scale firmly in place. Knowing the target answer is 14.70 cm, it's tempting to shift the scale so the 7th mm division lines up more favorably with your particular pencil mark.

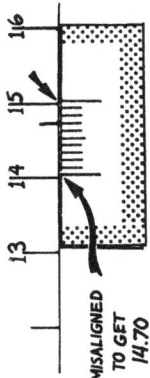

MISALIGNED TO GET 14.70

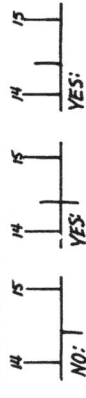

Double check answers that seem too perfect. Even expert estimators can err a tenth mm or two. If you draw up from the bottom but stop at the line, you'll later cover this pencil mark when you align the mm scale with the ruler. Be sure, therefore, to continue across the line, or draw down from above.

Materials
☐ Scissors

(TO) practice estimating between centimeter intervals. To check your accuracy with a millimeter scale.

NAME: _____
CLASS: _____

More Metrics ( )8

A SECOND LOOK

1 Start with a *sharp* pencil.

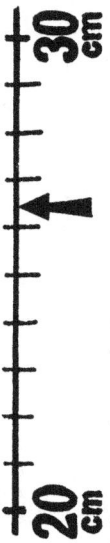

2 Cross this line at 14.7 cm. Use a thin pencil mark.

"14.7 cm"

14 . 7 | 2

(answers will vary)

3 Cut out this gray box. Fold it on the center line to make a mm scale.

THEN FOLD

4 Line up your cut-out mm scale so it fits *exactly* between 14 and 15 cm above . . . then estimate the position of your pencil mark to the *second decimal place* . . .

5 Find the difference. How close did you come to the 14.70 mark?

14.72 cm
−14.70 cm
_____
.02 cm

(answers will vary)

6 Fill in the table. *Finish one whole line before starting the next.*

MARK EACH MEASURE. Estimate . . . *don't use the cut-out scale.*	CHECK YOUR ACCURACY *with the mm scale to two decimal places.*	FIND THE DIFFERENCE. *How close did you come?*
	EXAMPLE 12.84 cm	EXAMPLE .04 cm
12.8 cm	cm: 12 13 14 15 16	
13.1 cm	12 13 14 15 16	
14.3 cm	12 13 14 15 16	ANSWERS
15.4 cm	12 13 14 15 16	WILL
14.6 cm	12 13 14 15 16	VARY
13.2 cm	12 13 14 15 16	
12.7 cm	12 13 14 15 16	

7 How do you make a ruler more accurate?

Use smaller subdivisions.

Is it possible to make a ruler so accurate that each measure is exact? Explain.

No. There is always uncertainty, even between the smallest subdivisions.

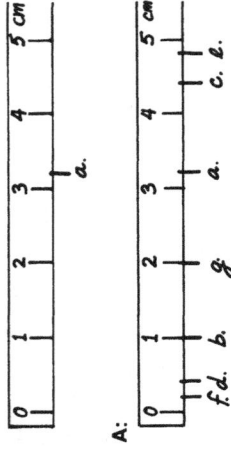

Teaching Notes 9

Discussion

If your class needs the practice, try this simple exercise. Select some measure — in millimeters, centimeters or meters — within the range of your computer can:

   0 - 160 mm
   0 - 16 cm
   0 - .16 m

Write the measure on your blackboard, asking students to dial it on their computer cans. Then circulate around your class to check answers. Start with easy measures (50 mm, 12 cm, .10 m). Work up to harder ones (14.6 mm, .09 cm, 106 m).

Evaluation

Q: Show where each measurement hits the ruler. Draw a thin pencil mark and letter it as shown in the example.

a. 32 mm   c. 44 mm   e. 4.8 cm
b. 10 mm   d. .4 cm   f. .002 m
                          g. .02 m

A:

Materials

- ☐ Scissors.
- ☐ A soda can.
- ☐ Clear tape.
- ☐ Masking tape.
- ☐ Thread.
- ☐ A penny.

---

2. Tape only where the two ruler ends come together. By keeping the scale itself tape-free, there will be no rough edges to snag the thread as you dial measurements in step 9.

4-7. Don't center the tape over the rivet. This causes the thread to emerge from underneath at a position that is very much off center.

WRONG:

Instead, place the *edge* of the tape across the rivet. And cover the hole. This keeps the penny from falling into the can and possibly getting sticky.

If properly centered, you can suspend the can by its thread and it will hang straight down.

RIGHT: **WRONG:**

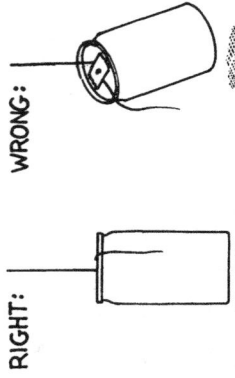

8. Always use the computer can on a level surface. Students with slanting desks may need to use the floor. Don't pick up the can after you dial a number. This allows the thread to remain in a fixed vertical position, held with hairline accuracy by the weight of the penny.

9. Students should double-check that each answer is correct before they write it down. They can work with someone else who also has a computer can. Or they can work alone, asking the teacher to check each answer.

PARALLAX

Clamp one end of a soda straw in the jaws of a clothespin. Tape the "wings" of your clothespin securely to a blackboard or wall.

Push a paper clip over the end of this straw, sliding it to about the middle. Slide a second straw *over* the whole paper clip so the 2 straws join to form a large "T".

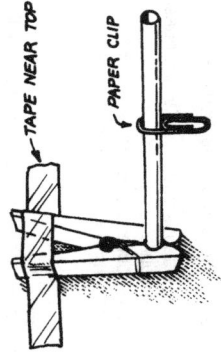

TAPE NEAR TOP
PAPER CLIP

(TO) accurately locate millimeters, centimeters and meters with a hairline on a metric scale.

NAME: _____

CLASS: _____   More Metrics ( )

DIAL A MEASURE

STRIP A 0 1 2 3 4 5 6 7 8 9 10 11 12 13 14 15 cm.

1 Cut out strip A at the side of this worksheet.

2 Wrap it around a pop can —just like a head band— and clear-tape the ends. Keep "A" upright.
STAY BELOW THE CURVE
PULL TIGHT AND EVEN!
TAPE ONCE.

3 Bend the pull-tab back and forth until it breaks off.

4 Cover the hole and *half* of the center rivet with masking tape.
Keep tape OFF THE RIM.
RIVET

5 Drape thread over the top so it crosses the center rivet, and tape.
THREAD CROSSES RIVET
TOUCHES TABLE.

6 Stick a second piece of masking tape exactly over the first.
THREAD IS "SANDWICHED" HERE

7 Tape a penny to the thread so it hangs just off the table. Trim both ends.

8 You can "dial" any measure just by pushing the thread around the rim.
TRY IT!
A COMPUTER CAN!

9 Work with a friend who also has a computer can. Follow these three steps.

1 DIAL each measure below.
DO 5.5 FIRST!

2 COMPARE your answer with your friend's.
CHECK!
5.5?

3 RECORD your answer on the scale below.

RECORD answers here:

a. 5.5 cm	e. 35 mm	i. .023 m	m. 83 mm
b. 11.1 cm	f. 134 mm	j. 6.4 cm	n. .117 m
c. 0.9 cm	g. .100 m	k. 0.1 cm	o. 46 mm
d. 70 mm	h. .140 m	l. 15.0 cm	SAVE YOUR CAN!

TOPS LEARNING SYSTEMS

Copy this scale. Tape it to your blackboard or wall directly behind the vertical straw so the scale and straw line up at the zero position.

-4 -2 0 +2 +4

Ask students to read the scale from where they sit. Start by calling on those at the extreme left. Proceed across the room, always moving right, until you reach the other side. Notice how the readings change.

The apparent movement of the straw across the scale is called parallax. Consider these questions as you discuss this concept:

● Where is parallax the most extreme, and where is it reduced to zero? (It's extreme at each side and zero in the exact middle.)

● If you move the vertical straw closer to the scale (or farther away), how does this affect the parallax? (It decreases as the straw approaches the scale and increases as it moves farther away.)

● To measure accurately, you need to keep parallax reduced to a minimum. Name 2 ways to do this. (Keep the hairline near the scale. Keep your viewing eye squarely on center, behind the hairline.)

3. Tape just the ends of scale B to the can, as you did before. This insures snag-free motion of the thread over the entire scale.

It's not necessary to align B with A. These 2 scales will be used independently as students play Agree/Disagree.

4. Since only one computer can is needed per student pair, only half your class needs to complete steps 1–4. Everyone should play the game, however, and all should give a thoughtful answer to question 7.

5. The first time they play, students will likely attempt to use both scales. Emphasize that the directions read "A scale only."

Once the hairline has settled on some random measure, don't move the can. Read the hairline where it comes to rest, lowering your head to get an eye-level view of the scale.

Agree/Disagree is a cooperative game. If both players write separate answers that agree, they advance 1 space. For example, 11.2 cm agrees with 11.3 cm because the answers differ by no more than a tenth. Answers with a difference greater than a tenth are wrong: slide the paper clip back 2 spaces, but no lower than zero. It's easy to win on the A scale because the estimated interval is a whole cm wide.

6. The B scale is more difficult to play because the estimated interval is only a mm wide. Both players have to read the hairline with much greater accuracy in order to agree. For example,

11.26 cm agrees with 11.27 cm, but not 11.28 cm. Answers with a difference greater than a hundreth are wrong: slide back 2 spaces!

Students who understand parallax (see box above) can probably win on the B scale. You can make the game easier, but less challenging, if you relax the standard of uncertainty from 1 digit to 2.

Evaluation

Q: Carlos and Sara each use one of the rulers below to measure the length of a toothpick. Carlos correctly measures 4.26 cm while Sara gets 4.3 cm. Which ruler did each student use? Explain how you know.

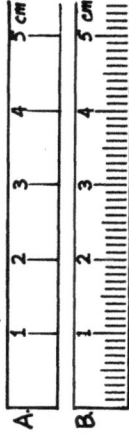

A: Carlos used ruler B because he estimated between mm intervals to the nearest .01 cm. Sara used ruler A because she estimated between cm intervals to the nearest .1 cm.

Materials

☐ Scissors.
☐ A computer can from step 9. Because students share this, only half your class will need one.
☐ Clear tape.
☐ A paper clip.

NAME: _____

CLASS: _____

(TO) agree with a friend, within acceptable limits of uncertainty, where the hairline crosses a scale.

More Metrics

AGREE / DISAGREE

1 Cut out strip "B" at the side of this worksheet.

2 Set the penny on top of your can, out of the way.

3 Wrap strip "B" around the can, just below A. Tape the ends together. *Tape just once!*

4 Find a friend to play Agree/Disagree. You'll need only one measuring can. *LET'S USE MINE!*

5 Play Agree/Disagree on the "A" band only. Follow these four rules. *IGNORE 'B' FOR THE FIRST GAME.*

1 SLIDE THE HAIRLINE to any chance position.

2 WRITE YOUR MEASURES on separate paper.

3 COMPARE YOUR ANSWER with your friend's.

4 ... UP 1 IF YOU AGREE: Uncertain digit off by no more than one.

... DOWN 2 IF YOU DISAGREE: Uncertain digit off by more than one, or certain digits disagree.

6 Now play Agree/Disagree on the "B" scale only. *Ignore 'A' this time.*

7 MOVE A PAPER CLIP ON THIS SCORE CHART . . . *Mark your place with a paper clip.*

SCORE CHART
WIN
5
4
3
2
1
START

Which scale is hardest to play, "A" or "B"? Why?

B. The uncertainty in the last digit increases as the divisions get smaller.

STRIP B 0 1 2 3 4 5 6 7 8 9 10 11 12 13 14 15 cm.

TOPS LEARNING SYSTEMS

Teaching Notes 11

We all have a tendency to make things come out even. It's easy to do. Just shift the ruler a hair this way or that. So be alert for zero bias, especially in the right column of numbers.

Students can check the consistency of their own answers before bringing their worksheets to you for a check point. Rounding off their measurements on the right to the nearest tenth, they should come to within .1 cm of the measurements on the left. Answers that fail this internal test should be remeasured.

Evaluation

A: a. Find the length of each stick. Be sure to estimate the last digit.

x = ...
y = ...

b. Which side of the ruler is more accurate? Why?

A: a. x = 3.4 cm
y = 4.48 cm

b. Side y is more accurate because it is subdivided into smaller millimeter intervals. This allows you to estimate between the spaces to the nearest tenth mm (.01 cm).

Materials

☐ Scissors

1-2. In this activity, students use rulers for the first time. Beginners may not recognize that many rulers, like the cut-outs below, have scales that are recessed away from the ends. Those who make the ruler's physical end their starting point, instead of the zero mark on the scale, will record line measures that are consistently too short.

INCORRECT:

CORRECT:

We've intentionally exaggerated this end space. We hope our wider gap makes it obvious that the scale's end and ruler's end are not congruent.

Ask students to preserve this space as they cut out each ruler. Don't trim it back to the zero mark. This will blur the distinction between the physical ruler and its scale, and likely impair the ruler's accuracy should the cut be imprecise. While cutting, be sure to leave the bottom scale straight and even. If gaps appear due to ragged cuts you can't bring the ruler close to a line without covering some parts of it.

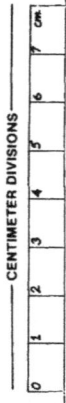

We have provided sample answers at the top of each column so your students will continue the pattern — estimating tenths on the left and hundredths on the right. Remember to write units, of course. The unit gives each number its meaning.

NAME:

CLASS:

More Metrics (   )11

LINE MEASURE

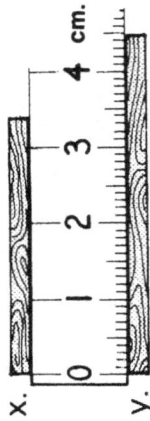

ESTIMATE THE LAST DIGIT— DON'T ROUND IT OFF!

1 Carefully cut out the *cm* ruler at the bottom of this page. Measure each line, then write your answer in the box.

WRITE YOUR UNITS ↓

a.	**2.6 cm**
b.	**1.2 cm**
c.	**5.1 cm**
d.	**4.3 cm**
e.	**7.7 cm**
f.	**0.9 cm**
g.	**5.3 cm**
h.	**2.0 cm**
i.	**0.4 cm**
j.	**6.6 cm**

a. _____
b. _____
c. _____
d. _____
e. _____
f. _____
g. _____
h. _____
i. _____
j. _____

2 Now cut out the ruler with mm divisions. Carefully re-measure each line and write your answer in the box.

WRITE YOUR UNITS ↓

a.	**2.62 cm**
b.	**1.24 cm**
c.	**5.09 cm**
d.	**4.31 cm**
e.	**7.73 cm**
f.	**0.89 cm**
g.	**5.30 cm**
h.	**2.02 cm**
i.	**0.39 cm**
j.	**6.56 cm**

3 Which ruler is more accurate — the centimeter one or the millimeter one? Why?

The millimeter ruler is more accurate because it has smaller subdivisions.

CENTIMETER DIVISIONS

MILLIMETER DIVISIONS

TOPS LEARNING SYSTEMS

(TO) practice measuring physical objects with a meter tape. To estimate the last digit.

METER MEASURE

1 Carefully cut around the outside of this gray box.

2 Now cut your gray box into 5 long strips. Tape these strips in order — 20, 40, 60, 80 — to make a meter tape.

LINE UP WITH THE STRAIGHT EDGE OF YOUR MARBLE.

3 Write your name on the back of your tape:

4 Find the length of each object in centimeters. Measure accurately — don't round off.

A 3x5 INDEX CARD:

diagonal: **14.8 cm**
width: w **7.6 cm**
length: l **12.7 cm**

A PENNY:

width: w **0.1 cm**
diameter: d **1.9 cm**

THIS WORKSHEET:

diagonal: **35.3 cm**
length: l **27.8 cm**
width: w **21.6 cm**

A SODA CAN:

circumference: c **20.7 cm**
height: l **12.2 cm**
diameter: d **6.5 cm**

5 Underline the *certain* part of each measurement above. Check with a friend to make sure your *certain* figures agree.

Do all the figures in each measure agree? Explain.

No. The last figure is sometimes different because it is uncertain.

SAVE YOUR METER TAPE

CHECK

TOPS LEARNING SYSTEMS

1-2. To make an accurate meter tape, cut carefully along the lines. Tape the 5 strips together so the ends just touch. Avoid overlaps that make the ruler too short, and gaps that make the ruler too long. If it is well pieced together, your tape should measure close to a standard meter.

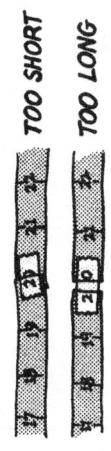

TOO SHORT

TOO LONG

Tapes that are significantly longer or shorter may have been distorted by your photocopier. This should not affect your results other than to make your answers for step 4 slightly higher or lower than those in our answer key.

4. We calibrated the meter tape in centimeters to give your students practical experience in estimating. Unlike millimeter intervals, centimeters are large enough to make estimating relatively easy. Moreover, few measurements can be expected to come out even.

Under no circumstances should students round off to the nearest whole centimeter. Rather, they must estimate between each interval to the nearest tenth centimeter.

Assuming little or no photocopy distortion, measurements below 20 cm should not vary from this key by more than a few tenths of a centimeter. We chose objects without size variations to keep these measurements uniform. Above 20 cm, expect more variations. The meter tapes themselves may vary, depending how one segment is taped to the next. Moreover, even slight photocopy distortions will add up to a significant cumulative error over longer distances.

Extension

You can verify two mathematical relationships dating back to the Greeks. Use measurements from your rectangles to illustrate the Pythagorean theorem.

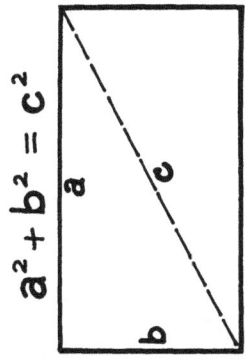

$$a^2 + b^2 = c^2$$

Use measurements from your cylinders to calculate pi — the ratio of the circumference to the diameter of any circle.

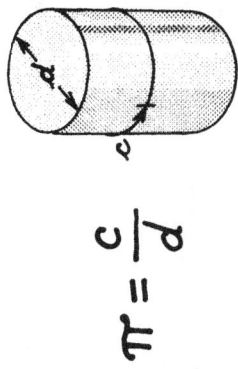

$$\pi = \frac{c}{d}$$

If you find that pi is slightly larger than 3.14, question the value of your diameter. Many cans round in at the bottom and slope in at the top. If you remeasure to the extreme outside circumference, you'll find the diameter slightly larger than originally estimated.

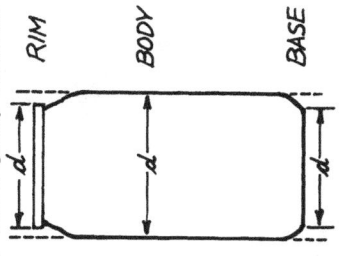

RIM
BODY
BASE

Evaluation

Q: Get a paper clip and a nickel. Measure each by laying them on this ruler.

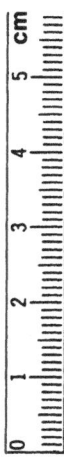

paper clip = ...
nickel = ...

Circle the uncertain part of each measure.

A: paper clip = 3.2̲0 cm
nickel = 2.1̲3 cm

(Although nickels have a standard size, your paper clips may differ from ours.)

Materials

☐ Scissors.
☐ Clear tape
☐ A 3 x 5 index card.
☐ A penny.
☐ A soda can.

Teaching Notes 12

1. Height is a dimension that is familiar to all. Spans and fathoms may require additional explanation. Ask your class to stand and stretch out their arms as far as possible: this is a fathom. Now separate thumb from little finger as far as possible: this is a span. Emphasize "as far as possible". The mathematical relationships in step 2 work best if students stretch to full extensions.

Your students will need to work in pairs so they can help each other hold the meter tape and read the scale. Each measure, of course, should be estimated between centimeter intervals to the nearest tenth. Units must accompany each answer.

Circumferences can be measured sitting or standing most anywhere in the room. Heights, fathoms and spans, however, are much easier to measure with pencil marks against a wall. Designate a special measuring area in your room. Corners work nicely. Attach butcher paper to the walls with masking tape to protect paint finishes.

3. Ask someone with large hands to measure the length of your blackboard in spans. Follow up with another student who has small hands. Ask why the first measure is shorter than the second. The blackboard is certainly not changing size. Discuss the idea of a measuring standard — a unit that is the same for everyone.

Evaluation

Q: Two students walk the length of a field to measure its length. One finds its length to be 290 paces while the other gets 313 paces. Explain why these students get different answers.

A: The student who measured out 313 paces has smaller steps than the student who measured out 290 paces. To get the same answer (within acceptable limits of error) they need to use a better standard for measure, a meter tape, for instance.

Materials

☐ A meter tape from activity 12.
☐ Butcher paper and masking tape (optional).
See note 1 paragraph 3 above.

BALANCING DEMONSTRATION

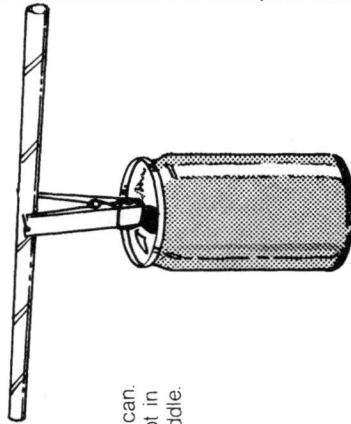

Stick a pin through the center of a straw near its top curve.

Clamp a clothespin to the pull-tab on a soda can. Rest the pin on the "ears" so the straw can pivot in the middle.

Demonstration

1. Show your class how to balance the straw beam with a rider. Stick a piece of tape lightly to the straw, then move it left or right to find the level position.

2. Challenge your class to balance the straw upside-down with the pin near the bottom. (The slightest breeze in your class makes this an impossible task.) Point out that a microbalance won't balance upside-down, either.

3. Compare the balancing properties of your straw with another poked very near its center.

RIDER

A: PIN NEAR TOP   B: PIN NEAR CENTER

(TO) discover basic body proportions by making accurate measurements with a meter tape. To appreciate that body measure is not standard.

NAME: _____

CLASS: _____   More Metrics ( )13

BODY MEASURE

1 Use your meter tape to find how you size up: *Estimate between lines, don't round off.*

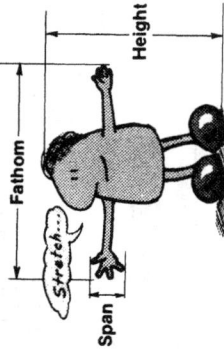

Height
Fathom
Stretch...
Span

WRITE THE UNITS!

Forehead
Neck
Wrist
Waist

CIRCUMFERENCES:

Forehead: ___ varied ___   Waist: ___ varied ___

Neck: ___ answers ___   Wrist: ___ answers ___

LENGTHS:

Height: _____

Fathom: ___ varied ___

Span: ___ answers ___

2 A very famous scientist* found that 8 times your span equals both your height and your fathom:

*Leonardo da Vinci

8 spans = 1 fathom = 1 height.

Use math to show how closely you fit this rule.

Varied answers. Here is a sample calculation.

height = 135.2 cm
fathom = 131.1 cm
span = 17.0 cm
8 spans = 8 x 17.0 cm = 136.0 cm

136.0 cm ≈ 131.1 cm ≈ 135.2 cm

3 Is it better to measure in spans or in meters? Explain.

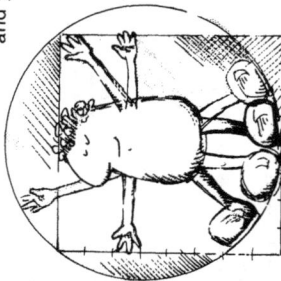

WHICH?

METER TAPE

SPAN

It's better to measure in meters. A span changes with body size, but a meter is standard — the same for everyone.

TOPS LEARNING SYSTEMS

(TO) improvise a single-arm balance beam, accurate to a milligram.

NAME: _____ CLASS: _____   More Metrics ( )14

BUILD A MICROBALANCE

1. Fold a 3x5 card in half the short way (like a tent). Write your name at the bottom.

2. Using a penny, draw a circle at the center of the fold, then cut a square around the circle. DRAW CIRCLE... CUT SQUARE

3. Unfold the square and tape it across the bottom of the tent.

4. Make a 1 cm snip in the end of a straw... 1cm CUT END ...Squash-fit the cut end inside another straw... ...Push them together to overlap 8 cm. 8 cm

5. Poke your double straw through the tent so the seam rests in the middle... SEAM ...Stick a pin through both straws exactly here. Near the top of the curve

6. Stick tape over this spoon handle where shown. Carefully cut out around the dashed line. FIRST TAPE / THEN CUT

7. Cut to the middle dot along the dotted line. Overlap the edges so the arrows come together, then tape. HANDLE / STICKY TAB

8. Curl the handle up so it slides into the short end of your straws. SLIDE IN

9. Push the spoon in all the way to its bowl ... then bend it level. PUSH IN TO BOWL / ROLL UPWARD

10. Fold some tape so one end is sticky ... and stick it next to the spoon. TAPE "RIDER"

11. Trim back the long end one mm at a time, until the spoon just tips down. OK—enough!

12. Move the rider to make the beam balance level.

TOPS LEARNING SYSTEMS

Students should observe that A is more stable, teetering back and forth across its equilibrium position quite rapidly until it comes to rest. Straw B is more sensitive, drifting off center at the slightest breeze.

Summarize with a blackboard diagram like this.

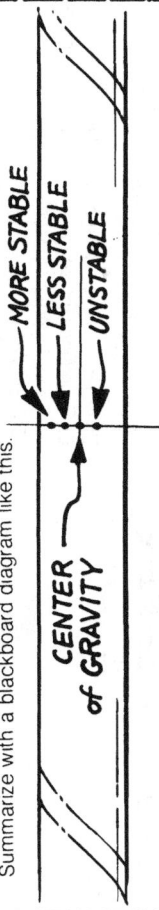

MORE STABLE
LESS STABLE
UNSTABLE
CENTER of GRAVITY

The pivot must be above the center of gravity to balance the straw at all. Any position you choose above this center is a trade-off between stability and sensitivity.

In step 5 below, we opted for stability. Choose sensitivity and you'll have to shut off the air conditioner (or heat) and close all the windows, just to keep the air calm enough for your balance to settle down.

4-5. The pivot pin divides the straw beam into a long and a short side.

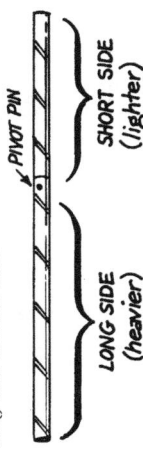

PIVOT PIN
LONG SIDE (heavier)   SHORT SIDE (lighter)

This allows you to add a weighing spoon (step 10) and a tape rider (step 11) without tipping the beam. The spoon should tip only when you trim back the long side (step 11); but not so much that it can't be centered by moving the tape rider back towards the pivot (step 12).

For average-length straws in the 16-22 cm range, an 8 cm overlap seems to create just the right disparity between the long and short sides. Longer straws require an extra cm of overlap, while shorter straws require less.

Be sure to build your own microbalance first, to test out your particular brand of straws. If the beam tips prematurely in step 10, increase the overlap. If you need to trim off too much straw to tip the spoon down in step 11, decrease the overlap. Then correct your worksheet original from 8 cm of overlap to your new figure. Be sure to do this before you duplicate class copies.

5. This pin placement serves 4 important functions. First it passes through both straws. This locks them together so they can no longer slide, one inside the other. Second, the pin is located just 1 cm inside the overlapping edge. This keeps the long and short ends of the straw at their proper lengths. Third, the pin pokes through the straw somewhere near the top of the curve. This increases the stability of the beam, making it easier to balance. (See demonstration above.)

Finally, the pin functions like a key. It actually locks the balance beam inside the tent window. You can't pull the beam back out of its tent without first shifting the pin key to one side.

6. Stick tape to the handle before you cut out the spoon. This enables the curled handle to fit more securely inside the straw (step 8). To keep the spoon as light as possible, use no more tape than necessary.

8. Curl the handle upward, as shown. A downward curl causes the spoon to droop.

10. Stick the rider very lightly to the straw so you can reposition it as necessary (step 12).

11. A dull pair of scissors will chew, rather than cut, a tough plastic straw. If your scissors aren't reasonably sharp, you can substitute a finger nail cutter in this step.

Evaluation

Is the pivot pin pushed through near the top edge of the straw? Is the spoon turned level relative to this upright position and pushed all the way in? Does the beam swing freely, returning to a balanced level position?

Materials

☐ A 3 x 5 index card.
☐ A penny.
☐ Scissors. They should be sharp enough to cut plastic straws. See note 11.
☐ Plastic soda straws.
☐ A tape rider from activity 12.
☐ A meter stick from activity 12.
☐ A straight pin. Long ones work best. (If you buy steel ones, they can also be used in our TOPS modules on Magnetism.)

Teaching Notes 14

c. **Without moving the slider or rider, replace your original object with something else.** In this activity it will be another seed, staple or paper clip. The rise or fall of the beam indicates whether the second object is heavier or lighter than the first.

Later, in activity 16, you'll add milligram masses to determine the actual mass of the object.

9. Among seeds of the same kind there is considerable mass variation. But not so much that it affects this ordering. Even the lightest pinto bean, for example, will still outweigh the heaviest popcorn.

10. You'll put much more in this cup besides straw sliders. In the next activity you'll add a whole array of milligram masses, plus tweezers.

Evaluation

Q: Two popcorn seeds look exactly alike. How would you use **only** a micro-balance (no mg masses) to prove that one is heavier?

A: Follow these three steps:

1. Center the microbalance by moving the tape rider right or left until the straw beam rests level.

2. Put 1 popcorn kernel in the center of the weighing spoon. Counterbalance its mass with a straw slider on the other side, shifting it until the straw beam rests level again.

3. Without moving the slider, replace the popcorn seed in your spoon with its look-alike. The spoon will rise if it's lighter or fall if it's heavier than the original seed.

Make sure your students understand this difference: the tape *rider* centers an empty system, while the straw *slider* recenters the beam after you put an object in the weighing spoon.

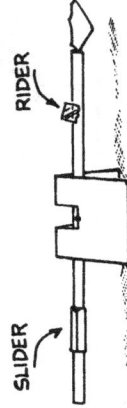

Materials

- ☐ A plastic soda straw.
- ☐ A meter tape from activity 12.
- ☐ Scissors.
- ☐ A microbalance from activity 14.
- ☐ Four kinds of seeds: pinto beans, popcorn, lentils and long-grain white rice.
- ☐ A staple.
- ☐ A paper clip.
- ☐ A paper or styrofoam cup.

To make accurate weight comparisons on your microbalance, you need to do 3 things:

a. **Center your empty balance with a tape rider.** Move it right or left, somewhere on the spoon side, until the beam balances level.

Do this each time you make a new weight determination. The spoon, especially, can cause an imbalance by shifting slightly inside the straw. So push it all the way in before you center. If the spoon seems extra loose, secure its bottom to the straw with a small piece of tape. Use tape sparingly. The lighter your balance, the more sensitive will be its response to small weight changes.

b. **Counterbalance the object you want to weigh with a straw slider.** First place the object on the white "bull's eye" in the center of the spoon. Then counterbalance its weight by slipping a straw slider of appropriate size over the opposite end. The three sliders from steps 1 and 2 should cover the whole weight range of objects you need to compare: the short slider for light objects; the medium slider for medium objects; the long slider for heavier objects.

Move the slider right or left until your system balances level. You can make fine lateral adjustments by twisting the slider as you go.

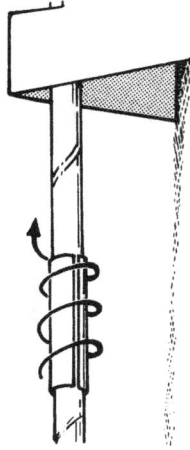

SLIDER   RIDER

(TO) learn to operate a microbalance. To compare the masses of small objects, ordering them from lightest to heaviest.

NAME: _____

CLASS: _____

More Metrics ( )15

SEEDS & SUCH

1 Cut a straw into 3 parts like this:

| 1cm | 5cm | — rest of straw — |

2 Slice each section open along its entire length.

3 Center your microbalance: move the tape rider so the straw balances level.

MOVE THE RIDER

4 Slide the *1 cm section* over your pencil, then onto the end of your balance.

THIS IS YOUR SLIDER

5 Put a rice grain in your weighing spoon. Move the *slider* until the beam balances.

RICE GRAIN

MOVE UNTIL LEVEL.

6 Does a rice grain weigh more than a staple? How does your balance tell you the answer?

No. When you replace the rice grain in the weighing spoon with the staple, the spoon end tilts down.

7 Which is heavier, a staple or a lentil? Use your balance and straw "sliders" to find out.

A lentil is heavier than a staple.

LONGER SLIDERS BALANCE HEAVIER OBJECTS.

8 Use your microbalance to find the *heavier* in each pair:

CIRCLE YOUR ANSWER

pinto bean / popcorn

paper clip / pinto bean

popcorn / lentil

9 Order these from lightest to heaviest:

- staple • popcorn • paper clip •
- rice grain • lentil • pinto bean

HEAVIEST GOES HERE

rice grain
staple
lentil
popcorn
pinto bean
paper clip

10 Put all 3 straw sliders in a cup. Save them to use again.

TOPS LEARNING SYSTEMS

(TO) make a set of milligram weights for the microbalance. To use them to find the masses of small objects.

MILLIGRAM MASSES

Name_____
This cup contains 3 SLIDERS, TWEEZERS, and 11 MASSES:

200 mg	20 mg	2 mg
200 mg	20 mg	2 mg
100 mg	10 mg	1 mg
50 mg	5 mg	+ extra squares

1 Write your name on this gray box. Then cut it out and tape it to the side of your cup of sliders.

2 Carefully cut out each paper mass and fold it so the number shows. Store all 12 in your cup.
CUT:
FOLD: NUMBERS MUST SHOW

3 Make a pair of tweezers:
Cut a straw into halves.
Bend one, and tape the elbow.
Cut the other in the middle at a slant...
...to make 2 POINTED ENDS.
Snip outside edges 1 cm.
Slide snipped ends INTO pointed pieces:
FINISHED TWEEZERS
TURN ENDS SO POINTS MEET.

4 Weigh each object in mgs: (Answers may vary:)

	lentil:	popcorn:	rice grain:	paper clip:	staple:	pinto bean:
MASS (mgs)	52 mg	137 mg	16 mg	495 mg	33 mg	284 mg

5 Do these results agree with activity #15? Explain.

Yes. The lightest object (a rice grain) weighs less than the next lightest object (a staple), right down the light-to-heavy list.
(Answers may vary)

6 How many rice grains make 1 gram (1000 mg)? Show your math.
(Answers may vary)

$$\frac{1{,}000 \text{ mg}}{16 \text{ mg/rice grain}} = 63 \text{ rice grains}$$

STORE SLIDERS, TWEEZERS & MASSES IN YOUR CUP

PAPER MASSES:

100 mg | 200 mg | 20 mg | 20 mg | 10 mg | 5 mg
200 mg | 200 mg | 50 mg
EXTRA SQUARES (Don't fold) | 2 | 1

TOPS LEARNING SYSTEMS

This activity enables students to see and feel milligrams first hand. We've reduced these units of mass to tiny squares of paper. All you have to do is copy them on the right kind of paper.

We recommend 20 pound bond, a medium weight paper used in many photocopiers. To find out if your paper has the correct weight, use it to copy the milligram masses on this worksheet (step 2), then weigh the objects in step 4 on a centered balance.

Allowing for minor size variations, your answers should agree with our answer key. If they are consistently high or low, by more than 10% try a different kind of copy paper. Use a heavier paper if your results are too high, and a lighter paper if your results are too low.

2. The more tightly you fold the milligram weights, the more compactly they rest in the weighing spoon.

It's best to cut and fold one mass before doing the next. Those who cut them all out together, before folding, risk losing the smaller ones among the paper clutter. Because the lighter mg masses are especially hard to keep track of, we've included spare squares. Use them to replace lost squares as necessary. Store all masses in your cup when not using them.

3. Tweezers are convenient for reaching into your weighing spoon to replace one small mg mass with another. But you can get along without them — especially poorly made ones. Simply tip all the weights out of your spoon, then pick up the ones you want, off the hard table surface, with your chubby fingers.

To make tweezers that work well, it's important to bend the elbow in one even crease.

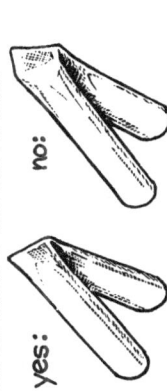

yes:

no:

The ends don't have to meet evenly. The points you add later compensate for any length difference.

Snip only the *outside* as illustrated. This enables the points to come together like pincers.

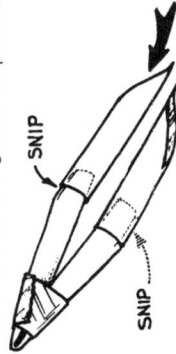

SNIP
SNIP

4. You already used your microbalance to compare relative masses (activity 15). Here the task is to find actual masses. Both procedures are identical except for one difference: replace the object you are weighing, not with another object, but with your mg masses instead. Just add your masses one by one to bring the beam to a level position. Then count what's in the spoon.

Add big mg masses to your spoon before you add the little ones. Save the small masses to make fine adjustments after you've focused on the approximate mass.

Extension

Challenge your students to weigh a penny in milligrams, then convert to grams.

They should find that a pre-1982 penny weighs about 3,000 mgs (3 grams). A post-1982 penny weighs 3,000 mgs less, about 2,500 mgs (2.5 grams).

Because the penny is so heavy, students must do some creative problem solving. They'll need to construct a slider at least 2 straws long, just to counterbalance the penny. Then they'll have to use some other material much heavier than paper, to counterbalance the slider. The spoon is too small to hold several thousand milligrams of paper weights.

We suggest weighing the penny with clay. Once they find an equivalent mass in clay, they can break it into pieces that are small enough to weigh with mg paper masses.

Evaluation

Q: Suppose you wish to know the mass of a raisin. Tell how you would use your microbalance to find the answer.

A: Follow these three steps:
1. Center the microbalance by moving the tape rider right or left until the straw beam rests level.
2. Put the raisin in the center of the weighing spoon. Counterbalance its mass with a straw slider on the other side, shifting it until the straw beam rests level again.
3. Without moving the slider, replace the raisin with your milligram masses. Add large ones, then small ones until the straw beam rests level yet again. The mass of the masses equals the sum of the masses in your weighing spoon.

Materials

☐ A microbalance plus its cup of sliders from activity 15.
☐ Scissors.
☐ Clear tape.
☐ A plastic soda straw.
☐ A meter tape from activity 12.
☐ Four kinds of seeds: pinto beans, popcorn, lentils and long-grain white rice.
☐ A staple and paper clip.

Teaching Notes 16

YES:

NO:

Weight is a measure of gravitational force. It changes as gravity changes. You weigh less on the moon than on Earth, but more on Jupiter. In free space you weigh nothing at all.

Mass, by contrast, is a measure of the amount of matter in an object. Your body is made of the same stuff, regardless of where you take it. Your mass is constant on Earth, on the moon, on Jupiter or in free space.

Our language and popular culture make no distinction between mass and weight. They are inextricably linked by Earth's nearly constant gravitational field. Dieters, for example, dream about losing "weight" even though they really ought to call it "mass". They would not be satisfied to live on the moon at one-sixth their normal weight. No, their mass — the fat and flab — would look just as awful.

This TOPS module makes no distinction between mass and weight either, for 3 good reasons. First, as long as you restrict use of this module to planet Earth, mass and weight don't need to be differentiated: they grow and shrink proportionately. Second, because our language doesn't recognize this distinction, our sentences would only seem unclear and convoluted if we labored to maintain it. Third, earth-bound beginning science students may not be intellectually ready to comprehend the difference between mass and weight.

Perhaps your students are ready. Then talk about mass and weight within the context of space travel, as we have above. Or bring the concept even closer to home: how does it feel to ride in an elevator? Do you shrink and expand as you feel lighter or heavier? No, Your mass (amount of matter) doesn't change even though your weight (force exerted by gravity) does.

1. If you know the weight of 1 rice grain, then allowing for small variations, 2 rice grains weigh twice as much, 3 rice grains 3 times as much, and so on. You can take advantage of this relationship by weighing only the first rice grain, then completing the rest of the table mathematically. Or you can directly weigh each number of grains. Either method produces a good result. Allow your students to follow their own inclinations.

2. Stress these graphing techniques:
   a. Mark each coordinate pair with a single sharp point. Circle it for emphasis.
   b. Draw the best possible straight line among your points. Use a straight edge. Don't draw from point to point as if you were "connecting the dots".

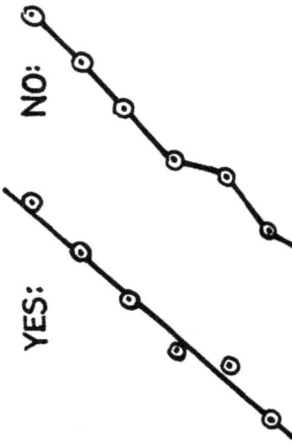

c. Don't draw through the circles. Preserve the clarity of each point so that others can verify you plotted it accurately.

5. Although the question only asks students to sketch an estimate, those who want to can easily verify their predictions. Simply make a data table similar to steps 1 or 3, then plot the ordered pairs on the graph in step 2.

Evaluation

Q: This graph shows how the mass of cheerios increases as you add more of them to a balance.

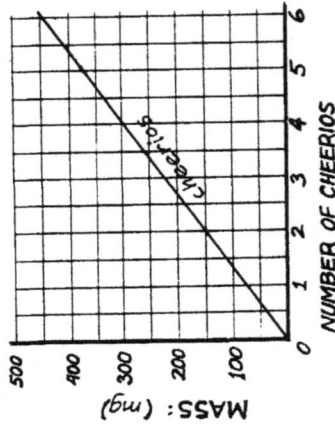

a. What is the mass of 2 cheerios? 5 cheerios?
b. Would you expect a "peanut" graph to be less steep or more steep? Explain.

A: a. 2 cheerios = 150 mg
5 cheerios = 375 mg
b. Peanuts would have a much steeper graph line (slope) because each one has far more mass, on average, than each cheerio.

Materials

☐ A microbalance plus its cup of accessories — sliders, weights and tweezers.
☐ Rice grains and lentils.
☐ A folded index card or other straight edge.

NAME:

CLASS:

More Metrics ( )17

(TO) graph how the mass of seeds increases in direct proportion to their numbers. To learn to draw and interpret graphs.

SEED GRAPHS

1 Weigh rice grains on your microbalance. Fill in this table.

HOW MANY	MASS (mgs)
0	0
1	16
4	63
10	156

2 Plot and circle each point on the graph. Draw the best possible straight line through your points and label it "rice grains."

DON'T DRAW INSIDE CIRCLES.

A FOLDED CARD MAKES A GOOD STRAIGHT EDGE.

GO TO BOX 3

3 Now fill in this table for lentils. Draw another graph line and label it "lentils."

HOW MANY	MASS (mgs)
0	0
1	52
2	110
4	218

4 Read your *graph* to estimate the mass of . . .

. . . 7 rice grains.	**110 mg**
. . . 3 lentils.	**163 mg**

Does your microbalance prove your graph results are correct? Explain.

CHECK YOUR ESTIMATES

Yes. Seven rice grains weigh about 110 mg — much lighter than 3 lentils.

5 Staples weigh about 30 mg each. Predict how a graph line for staples compares to rice and lentils.

DRAW AND LABEL 3 LINES.

TOPS LEARNING SYSTEMS

(TO) construct a dripper that dispenses uniform water drops. To confirm that a teaspoon holds 5 ml of liquid.

NAME: _____ CLASS: _____ More Metrics ( )18

WATER DOMES

1 Carefully cut out this "drip strip" along the dashed line.
PINHOLE

2 Tape your drip strip to a cup exactly as shown. The gray triangle must hang even with the bottom edge.
TAPE JUST ABOVE PINHOLE
BOTTOM EDGE

3 Cut away the top, just above the drip strip.
CUT CLEAR AROUND.

4 Cut off only half of the top of another cup.
AS SHORT AS THE FIRST ONE

5 Paper clip the short cup inside the tall one so the drip strip hangs in the opening.

6 Poke a pin through the drip strip hole and into the cup. Pull it back out.

7 Pour water into the top cup....
...To start it dripping, wet the paper on both sides with your fingers.
...To stop it dripping, push the pin back in the hole.

8 Balance a penny on a half clothespin. See how many drips it holds without spilling.

CATCH DROPS IN FREE FALL.
DRY PENNY AND CLOTHESPIN BETWEEN TRIALS.

RECORD 10 TRIALS
1 2 3 4 5 6 7 8 9 10

Answers should fall in the range of 12 to 15 drops.

9 A penny holds about 1 ml of water. Throw out trials you think are mistakes, then find an average.
AVERAGE DROPS IN 1 ml: **14 drops**

10 A level (not heaping) teaspoon holds 5 ml.
Predict how many drops this makes: $14 \times 5 = 70$ drops
Test your prediction: 70 drops.
A teaspoon holds roughly 70 drops.

11 Empty all water from your dripper. Put your name on it, and save it for the next activity.
NAME

TOPS LEARNING SYSTEMS

Over time, the rate of dripping tends to slow. You can speed the drips along by adding additional water to the top cup. This increases pressure at the pinhole. Or you can enlarge the hole slightly with your pin. One drip per second seems to be an optimum speed.

9. If you do it right, you can collect from 12 to 15 drops on the head of a penny. These are the large full-bodied variety, falling from fully saturated drip strips.

Evaluation

Q: If 20 drops from a drippy faucet fit on the head of a penny, how many fill a level teaspoon? Explain.

A: The water dome on a penny occupies about 1 ml. A level teaspoon holds about 5 ml, or 5 times as much. The teaspoon, therefore, holds 100 drops.

Materials
☐ Scissors.
☐ Tape.
☐ Paper or styrofoam cups, 6 oz. or more.
☐ A paper clip.
☐ A straight pin.
☐ A pitcher of water or sink with running water.
☐ A half clothespin.
☐ A penny.
☐ A teaspoon.

Teaching Notes 18

---

2. Position the drip strip exactly as shown: no higher, no lower. This insures enough overhanging paper to form a large, well-rounded drip, but not so much that you can't slide a penny underneath in step 8.

7. Water probably won't pass through the pinhole until you thoroughly wet the drip strip on both sides, from the pinhole on down. This wetness creates a wicking action. As drips fall from the end, additional water is pulled through the pinhole by capillary action.
   Water tends to drip in larger and larger drops as the paper strip becomes increasingly saturated.

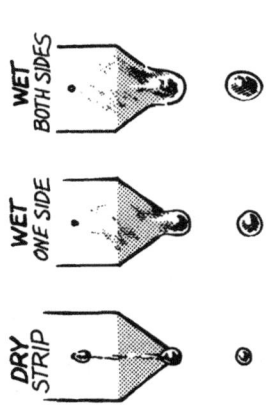

DRY STRIP   WET ONE SIDE   WET BOTH SIDES

By thoroughly wetting both sides of the paper beforehand, you eliminate this progressive drip growth. The drips start out large and remain large.

A big advantage with big drops is that they have volumes and masses that stay roughly uniform. You can measure just one on your microbalance (activity 19) to deduce how much water you can heap on the head of a penny.

8. Ten trials may seem excessive. Not really. Students love to build water domes and they need the practice. Here are some "don'ts" that drip catchers must learn to avoid in order to become expert!

a. Don't hold the penny too close to the drip. When this happens, water flows straight into the dome without dripping.
b. Don't forget to dry the penny and clothespin between trials. This keeps the water from spilling prematurely.
c. Don't tilt or shake the penny so the water dome collapses too early.
d. Don't fail to notice subtle dome spills that slip over the edge at eye-blinking speed.
e. Don't miscount. The drip that initiates the spill should not be included in your total.

Teaching Notes 19

All metric units derive from this humble metric water cube. A hundred placed end to end make 1 meter. A thousand stacked tightly together take up a volume of 1 liter. One of them weighs 1 gram.

You already know the number of drops contained in a water dome from activity 18. This activity provides the metric follow-up, asking you to determine the mass of this dome in mgs and its volume in mls.

1. Your first task is to find the mass of 1 drop. Your microbalance is the perfect tool for the job. To keep its weighing pan dry, you need to fashion a tiny foil dish.

Pressing the round foil firmly into the soft palm of your hand imparts a gentle curve to the dish that easily contains a water drop. Students may attempt to augment this curve by turning up the outside edges. Not only does this spoil its esthetic shape, but by narrowing its opening, you make it more difficult to capture a water drop inside (step 3).

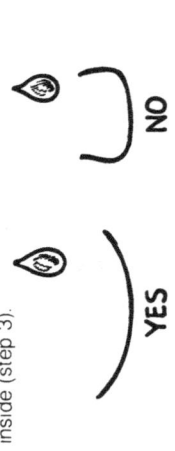

YES   **NO**

The microdish may lose its concave shape during the course of handling. You can renew its shape at any time by molding it again against the curve of your finger and the palm of your hand.

3. Thoroughly saturate both sides of the paper with water so the drips fall as large as possible. Do this before you collect one for weighing. This will stabilize the drip size, making it roughly equivalent to the size that formed the water domes in activity 18.

4. If your dripper is producing large full drops falling from a fully saturated strip, your 3 trials should be reasonably close. If your range is greater than 10 mg, experimental error probably scrambled the results.

6. Covering the cube with tape serves two purposes. First, you make it waterproof, at least for a short while. Second, you stiffen the sides of the cube so they don't bow out too severely when you add water. When this happens the centimeter water cube holds more than its alloted 1 ml.

Evaluation

Q: A drippy faucet has drops that weigh about 40 mg each. How many of these water drops make 1 ml?

A: 1 ml = 1 g = 1000 mg

$$\frac{1000\ mg}{1} \times \frac{1\ drop}{40\ mg} = 25\ drops$$

Metric Word Play!

Q: 2,000 mockingbirds = ?
10 cards = ?

2 kilomockingbirds
1 decacards

A:

1 ml :
STRAIGHT SIDES

1 ml plus :
BULGING SIDES

Materials

- [ ] A penny.
- [ ] Aluminum foil.
- [ ] Scissors.
- [ ] A microbalance plus its cup of accessories — sliders, weights and tweezers.
- [ ] A water dripper from activity 18.
- [ ] A pitcher of water or sink with running water.
- [ ] Clear tape.

NAME:

CLASS:

More Metrics (   ) 19

METRIC DRIPS

(TO) find the mass of a uniform water drop. To compare the volume of water a penny can hold with the volume contained in a centimeter cube.

1 Make a micro-dish: Press a penny into aluminum foil, then cut around the outline . . .

PRESS   CUT

. . . Curve your "dish" by pressing it into your hand.

2 Center your balance, then find the mass of your micro-dish.

HOW MANY MGS?

12 mg

3 Set up your dripper as before . . .

MINE

. . . Wet both sides of the paper so the drops flow fat and slow . . .

SLOPPY WET!

MINE

Catch just one on your micro-dish in free-fall.

Don't touch the drop before it falls.

4 Find the mass of this drop, then 2 others. Dry your dish and center your balance after each trial.

DROP:	1	2	3	AVERAGE:
DISH & DROP	89 mg	85 mg	90 mg	
-DISH ONLY	12 mg	12 mg	12 mg	
=DROP ONLY	77 mg	73 mg	78 mg	**76 mg**

5 Use your result from activity 18 to estimate the total mass of water a penny can hold.

WATER DOME:

HOW MANY MGS?

1 penny holds 14 drops.

$$14\ drops \times \frac{76\ mg}{drop} = 1064\ mg$$

6 Cover this cube pattern with tape. Cut it out and tape it together so it can hold water!

TAPE, THEN FIRST, THEN CUT IT!

1 cm³
1000 mg
1 ml
1 gm

7 Do you think your cube holds more than your penny, less than your penny, or about the same? Explain.

The cm cube holds 1000 mg of water. This is almost as much as the penny will hold (1064 mg).

8 Test your prediction. What do you find?

13 drops filled the cube about even with the top.

$$13\ drops \times \frac{76\ mg}{drop}$$
$$= 988\ mg$$
$$\approx 1000\ mg$$

TOPS LEARNING SYSTEMS

Teaching Notes (column)

5. If you choose a candy or nut that is heavier than 500 mg, your students will need to use a longer slider, perhaps a whole straw or two, as a counterbalance. Then they'll need to pool their mg masses together to counterbalance the slider.

As an alternative to using extra mg weights, provide clay. Once your students find an equivalent mass in clay, they can break it into pieces that are small enough to weigh with a single set of 500 mg masses.

Extension

Estimate the number of cheerios in a whole box. How many might it cost a penny? Both problems require long division. Provide a calculator.

$$\frac{gms}{1\ box} \times \frac{1\ cheerio}{gms} = \frac{cheerios}{box}$$

$$\frac{cheerios}{box} \times \frac{box}{pennies} = \frac{cheerios}{penny}$$

Evaluation

Q: A rice grain weighs 16 mg. Estimate how many grains fill a kilogram package.

A: $1\ kg \times \frac{1000g}{1\ kg} \times \frac{1000\ mg}{1\ g} \times \frac{1\ rice\ grain}{16\ mg}$

= 62,500 rice grains

Materials

□ A bottle of plain uncoated aspirin.
□ A microbalance plus its cup of accessories — sliders, weights, and tweezers.
□ Uniform treats such as candy, nuts, or raisins. See teaching notes 4 and 5 above. You'll need at least 1 open package for your students to experiment with, plus another unopened package to count and perhaps give away at the end of the lesson.

1. The instructions in this step are intentionally vague. If your class is large and your aspirin supply small, have students weigh the **same** tablet 3 times, then take an average. You can list all individual averages on your blackboard, then find a class super-average.

If your aspirin supply is plentiful, ask students to weigh 3 **different** tablets, then find an average for all 3.

The average (also called the mean) is easier to figure if you find deviations. Consider these two ways to find an average for the 3 trial values in the answer key:

hard way:
375
362
367
1104
mean = 1104/3
= 368

easy way:
362 + 13
362 + 0
362 + 5
       18
mean = 18/3
deviation = 6
mean = 362 + 6 = 368

3. Most aspirin tablets contain cornstarch as an added ingredient. This helps bind the tablet together.

4. What kind of a treat should you provide? Nuts, seeds or raisins are the most nutritious. Candy will likely be more popular. As you make a selection consider these criteria:

Size Appropriate? Jelly beans are too large for your microbalance. Salt grains are too small.

Number Appropriate? The package should not contain so many that it's difficult to count them at the end of the activity to find who estimated closest. Nor should it contain so few that it's easy to count them in an unopened package. If numbers are relatively small, the package should be opaque.

How Uniform? Regular-size M and M brand chocolate candies without peanuts are easy to estimate; irregular-size peanuts present more of an estimating challenge.

Net Weight? This must be printed on the outside of the package. Use the gram net weight or multiply ounces by 28.35.

As a non-eating alternative, weigh out exactly 10 grams of rice and store it in a labeled baggy.

Worksheet

(TO) use a microbalance to determine if aspirin are 100% pure. To estimate total quantity based on the mass of a few.

NAME:

CLASS:

More Metrics ( )20

CONSUMER SCIENCE

1 Weigh uncoated aspirin tablets 3 times. Find the *average mass*.

Trial 1	375 mgs
Trial 2	362 mg
Trial 3	367 mg
Average	368 mg

2 Read the label. Notice that each tablet contains 5 "grains" of aspirin.

One "grain" equals 65 mgs. How many mgs of aspirin are in your tablet?

$$\frac{65\ mg}{grain} \times 5\ grains = 325\ mg$$

3 Compare: is your tablet made from *all* aspirin (100% pure) or does it contain added ingredients? Explain.

The tablets contain 325 mg aspirin plus about 43 mg of something else.

4 Read the net weight (in grams) on a small package of *unopened* candies, nuts or raisins.

How many grams?

How many mgs?

depends on treat

5 Weigh just one piece on your microbalance.

...and a longer slider!

You may need to make another 500 mg mass from foil.

How many mgs?

depends on treat

6 Now use math to estimate how many fill the whole package. Whoever comes closest might win the whole package!

SHOW YOUR MATH!

$$\frac{net\ weight}{wt\ of\ average\ piece} = Number\ in\ package$$

HOW MANY GRAINS?

10 GRAMS RICE

HOW TO REMOVE WORKSHEETS

Perforated worksheets don't always work like they are supposed to.

This book is designed not only to make your science lessons run smoothly, but to make the worksheets pull out smoothly as well. Our pages are "perfect bound" in the same manner as single sheets of stationary are attached to a writing pad. You can remove worksheets from this book just like pulling sheets off a pad — well, almost.

We didn't want our book to shed leaves like a tree. So we ordered the perfect binding very strong. To remove the worksheets cleanly and quickly, be sure to follow one of these two special procedures.

One-at-a-time:

Start from the *back* of the book. *Pull* the top sheet off as illustrated. Don't tear. Proceed to the next until you remove all the worksheets.

The top sheet will probably be glued most securely to the binding. Sheets underneath should pull off more easily. Do not attempt to remove pages from the middle of the book *first*. This is often difficult to do (even on a scratch pad).

PULL GENTLY OUTWARD— LAST PAGE FIRST. DON'T TEAR!

Radical Surgery:

Place a sharp knife on this very page with the edge facing the binding. Close the book, and pull the knife through the binding to cleanly remove all the worksheets. Strip off the back cover and separate each page.

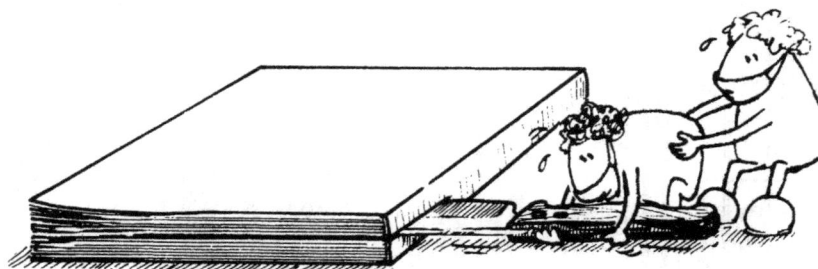

REPRODUCIBLE
STUDENT
ACTIVITY SHEETS

MILLI, CENTI, DECI & MORE

1 Cut off 7 strips of note-book paper along the blue lines.

ONE SPACE WIDE ↓

2 Fold one strip in half 4 times, then unfold it. This makes 16 equal parts.

16 parts— count them!

3 Cut off just 1/16 and

tape it here

Let's call this a MILLI.

LABEL YOURS "MILLI"

4 One *milli* is only one paper thick. Open a clothes-pin clip . . .

— 1 MILLI wide. —

paper layer

Ten *milli* make one *centi.* Now open your clip . . .

— 1 CENTI wide. —

paper layers

Ten *centi* make one deci. Now open your clip . . .

— 1 DECI wide. —

paper layers

5 Draw a line showing how far 1000 layers will reach.

USE YOUR DECI TO ESTIMATE.

6 Estimate how many sheets of paper you can stack from floor to ceiling. Write how you did it.

?

7 *Stack to the moon?*

If 1,000 papers stack about 8 cm, how many reach 4 cm?

(Half the distance)

How many sheets reach 4 meters?

(100 times farther)

How many sheets reach 4 kilometers?

(1,000 times farther)

How many sheets of paper would stack to the moon?

(100,000 times farther)

ABOUT 400,000 KM.

TOPS LEARNING SYSTEMS

KILOMETER RULER

1 Find 7 units of measure on this ruler. Write them in order here:

LONGER

⬆ each unit is ten times

SHORTER

one KILOMETER (km) 1,000 meters

100 meters

one HECTOMETER (hm)

10 meters

one DEKAMETER (dkm)

100 centi-meters.

one METER (m)

one CENTIMETER (cm)

one MILLIMETER (mm)

one DECIMETER (dm)

2

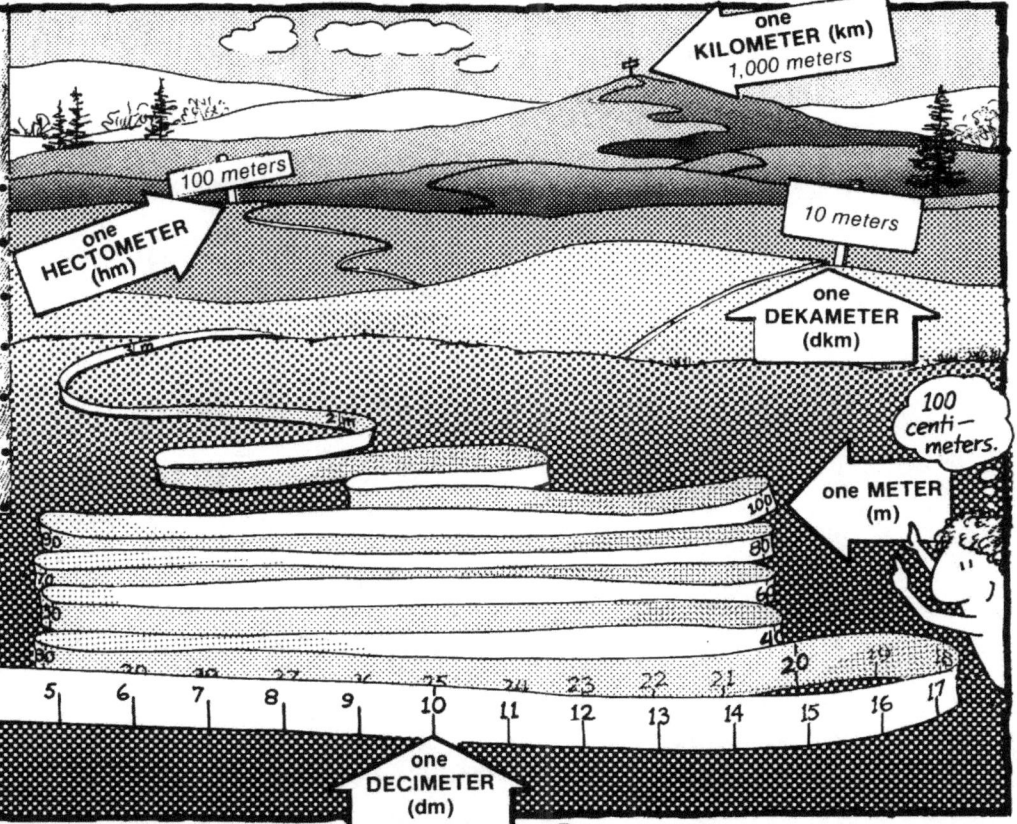

How many *millimeters* in . . .	How many *centimeters* in . . .	How many *meters* in . . .
1 cm?	50 mm?	100 cm?
5 cm?	35 mm?	50 cm?
2.5 cm?	5 mm?	20 cm?
7.8 cm?	1 mm?	19 cm?
70 cm?	250 mm?	10 cm?
1 dm?	1 dm?	3 cm?
3 dm?	9 dm?	1 cm?
1.6 dm?	.5 dm?	5 mm?
1 m?	1.2 dm?	1 mm?
4 m?	1 m?	300 mm?
1 dkm?	1 dkm?	1 km?
1 km?	1 hm?	5 km?

3 List the 7 stars here:

1 cm = 10 mm

These are used often! Memorize them!

TOPS LEARNING SYSTEMS

METRIC SQUARES

1 Get a sheet of METRIC SQUARES. Cut out all 42 squares *and* 3 labels.

2 Sort your squares into 3 labeled piles:

VOLUME	MASS	LENGTH
6 SQUARES	12 SQUARES	24 SQUARES

LITERS

GRAMS / MORE GRAMS

METERS

3 Now sort each pile into equal triplets—groups of 1 white, 1 grey, and 1 black that are equal.

VOLUME	MASS	LENGTH
(2 triplets)	(4 triplets)	(8 triplets)

TRIPLETS are always EQUAL

Record each set below!

4 Write each triplet in the correct space below.

V O L U M E

One liter is . . . . *as much as* . . . . and equals . . . .

One milliliter is . . . . and equals . . . .

M A S S

One kilogram is . . . . and equals . . . .

One gram is . . . . and equals . . . .

One gram is . . . . and equals . . . .

One milligram is . . . . and equals . . . .

L E N G T H

One kilometer is . . . . and equals . . . .

One meter is . . . . and equals . . . .

One meter is . . . . and equals . . . .

One meter is . . . . and equals . . . .

One centimeter is . . . . and equals . . . .

One centimeter is . . . . and equals . . . .

One millimeter is . . . . and equals . . . .

One millimeter is . . . . and equals . . . .

SAVE your metric squares!

TOPS LEARNING SYSTEMS

METRIC SQUARES

CUT ALL DASHED LINES

VOLUME		MASS		LENGTH	
ONE gram	As heavy as 2 raisins	.001 kg	ONE kilogram	As far as a 10 minute walk	.001 m
As heavy as a fruit fly	.001 g	ONE centimeter	As heavy as 2 paper clips	10 mm	ONE millimeter
1000 m	ONE liter	As wide as a doorway	.1 cm	ONE centimeter	As long as a pace
ONE milligram	As thin as a slice of bread	1000 mm	ONE gram	As much as 3 cans of pop	.001 km
As long as a flea	100 cm	ONE meter	As much as 1/5 teaspoon	.001 ℓ	ONE meter
.01 m	ONE kilometer	As thin as a dime	1000 g	ONE millimeter	As long as a monkey's tail
ONE meter	As wide as a fingernail	1000 mg	ONE milliliter	As heavy as a textbook	1000 ml

FACE-UP

1 Lay out a grid of masking tape on your desk so your metric squares can fit inside.

Like tic-tac-toe

2 Shuffle your deck of 42 cards to mix them well.

3 Place the top 10 cards *face up* on your grid.

Start with 2 cards in the middle...

...1 card everywhere else.

4 Search for 2 *equal* cards among the 9 on your grid.

Hmm — 2 equal cards?

DRAW PILE (face up)

. . . Cover each pair you find with 2 new cards from your draw pile.

Put them *face up* so they can form new pairs.

Equal!

5 Keep playing until . . . .

. . . *you win!* (all cards played)

Matched them all!

DRAW PILE GONE

. . . *you lose!*

(cards left in draw pile)

...stuck...

Teacher check

☐

6 Now try these games:
SOLITARY FACE-UP

I see a pair!

Repeat steps 2-5 on your own.

TEACHER ✔: ☐☐☐

COOPERATIVE FACE-UP

your turn...

...1 mg is as heavy as a flea.

Take turns finding 2 equal cards. Say each pair out loud.

TEACHER ✔: ☐☐☐

COMPETITIVE FACE-UP

100 cm = 1 meter!

RATS — now we're TIED!

Try to be first to call out pairs. Keep score.

TEACHER ✔: ☐☐☐

SAVE your squares.

TOPS LEARNING SYSTEMS

METRIC RUMMY

1 Cut a METRIC CARD HOLDER around the outside dashed line.

2 Fold it down, then up, like a fan. The words VOLUME, MASS and LENGTH all fold *down*.

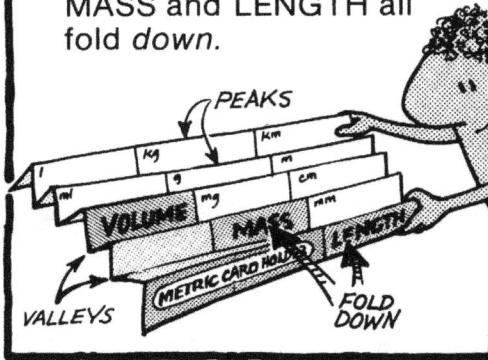

PEAKS

VOLUME MASS LENGTH

METRIC CARD HOLDER

VALLEYS

FOLD DOWN

3 Find a friend with a card holder and a deck of 42 shuffled cards.

Let's play metric rummy!

OK.

42 SHUFFLED CARDS

4 Deal 7 cards each, face down, and put them in your card holders . . .

This is my "hand."

. . . put the rest, face down, in a draw pile. Turn the top card up to start a discard row beside it.

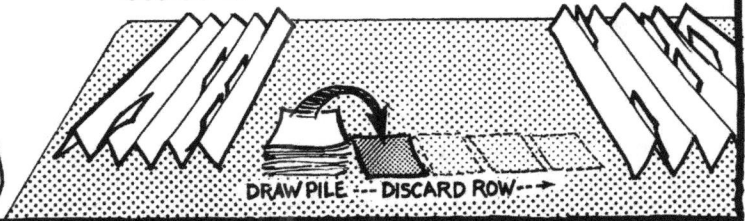

DRAW PILE --- DISCARD ROW--→

5 Take turns. Always follow 3 steps for each turn.

1 draw

2 match

3 discard

Draw 1 card from the draw pile *or* the discard row.

Take only the *newest* discard . . .

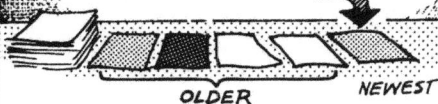

OLDER   NEWEST

. . . unless an older one forms an *instant* triplet in your hand.

COMPLETES A TRIPLET

TAKE ALL THESE, TOO

Match all the equal triplets you can.

WHITE = GRAY = BLACK

Lay these in front of you, reading each card aloud.

⬜ ... A meter ...

◻ ... As long as a monkey's tail...

◼ ... 100 cm!

... Those are right ...

Put one of your remaining cards at the *end* of the discard row.

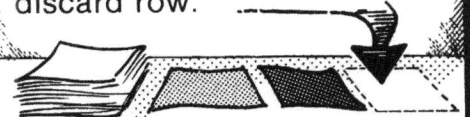

If you discard your *last card*, shout *"rummy"* and score!

RUMMY!

LAST CARD

SCORING
- ✪ Add 1 point for each card you played in a triplet.
- ✪ Subtract 1 pt. for each card left in your holder.
- ✪ *HIGHEST SCORE WINS!*

TOPS LEARNING SYSTEMS

LENGTH

mm

cm

m

km

MASS

valley

peak

valley

peak

mg

g

kg

VOLUME

ml

l

METRIC CARD HOLDER

TOPS LEARNING SYSTEMS

THE LAST DIGIT

1
Write the correct measure in each box. Always make the last digit a *zero* when the arrow points directly to a line.

example:

30.60 cm

ZERO means ON THE LINE.

cm: 30 31 32 33 34 35 36 37 38 39 40 41 42 43 44

h. c. i. a. b. e. d. j. f. g.

☐ TEACHER CHECK

| a. | b. | c. | d. |
| e. | f. | g. | h. | i. | j. |

2
These arrows point *between* lines. Imagine that each space is divided into ten parts, then *estimate* which tenth comes closest to the arrow.

example:

60.3 cm

I estimate about 3/10 of the way between.

cm: 60 61 62 63 64 65 66 67 68 69 70 71 72 73 7

a. g. i. f. c. d. b. j. e. h.

☐ TEACHER CHECK

| a. | b. | c. | d. |
| e. | f. | g. | h. | i. | j. |

3
Write each of these measures in 4 digits. Be careful about the last digit.

ESTIMATE if between the lines. Write ZERO if on the line.

TWEEN ON

cm: 41 42 43 44 45 46 47 48 49 50 51 52 53

g. c. h. f. d. e. i. j. b. a.

☐ TEACHER CHECK

| a. | b. | c. | d. |
| e. | f. | g. | h. | i. | j. |

TOPS LEARNING SYSTEMS

CERTAIN & UNCERTAIN

1 Write 2 possible measures for each arrow.

The .4 is estimated. It could be .5!

RULER A   50   51   52   53   54   55   56   57   58   59   60   61   62   cm.

example: ↑a.   ↑d.   ↑b.   ↑f.   ↑e.   ↑c.

a.	b.	c.	d.	e.	f.
50.4 cm					
50.5 cm					

51.1 is certain, but the last digit is uncertain.

RULER B   50   51   52   53   54   55   56   57   58   59   60   61   cm.

example: ↑g.   ↑j.   ↑h.   ↑k.   ↑i.   ↑l.

g.	h.	i.	j.	k.	l.
51.17 cm					
51.16 cm					

2 Good measurements have a *certain* part and an *uncertain* part. In step 1 above, *underline* the certain numbers, then *circle* what's uncertain.

50.④ cm

CERTAIN— for sure!   *UNCERTAIN— estimated!*

3 Which ruler is more accurate, A or B? Why?

Can you make a ruler that has no uncertainty? Explain:

4 These folks are having an argument.

50   51   52   53   54

It's 51.7!   *No! 51.8!*

Can both be right? Explain.

TOPS LEARNING SYSTEMS

A SECOND LOOK

1 Start with a *sharp* pencil.

2 Cross this line at 14.7 cm. Use a thin pencil mark.

14.7cm...

cm: 1|3   1|4   1|5   1|6

3 Cut out this gray box. Fold it on the center line to make a mm scale.

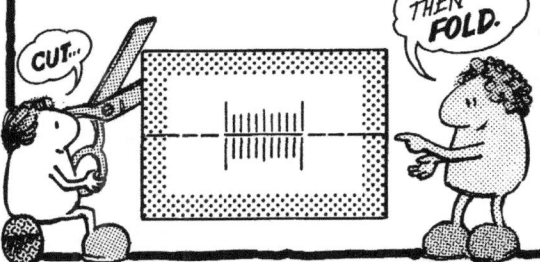

CUT...

THEN FOLD.

4 Line up your cut-out mm scale so it fits *exactly* betwen 14 and 15 cm above . . .

. . . then estimate the position of your pencil mark to the *second* decimal place . . .

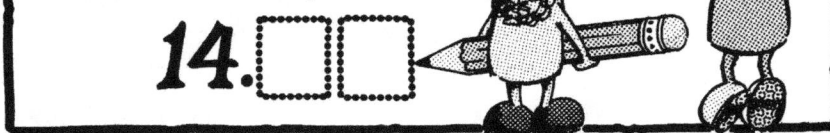

14.☐☐

5 Find the difference. How close did you come to the 14.70 mark?

6 Fill in the table. *Finish one whole line before starting the next.*

	MARK EACH MEASURE. Estimate . . . **don't** use the cut-out scale.	CHECK YOUR ACCURACY **with** the mm scale to **two** decimal places.	FIND THE DIFFERENCE. How close did you come?
12.8 cm	1|2 1|3 1|4 1|5 1|6		
13.1 cm	1|2 1|3 1|4 1|5 1|6		
14.3 cm	1|2 1|3 1|4 1|5 1|6		
15.4 cm	1|2 1|3 1|4 1|5 1|6		
14.6 cm	1|2 1|3 1|4 1|5 1|6		
13.2 cm	1|2 1|3 1|4 1|5 1|6		
12.7 cm	1|2 1|3 1|4 1|5 1|6		

7 How do you make a ruler more accurate?

Is it possible to make a ruler so accurate that each measure is exact? Explain.

TOPS LEARNING SYSTEMS

DIAL A MEASURE

1 Cut out strip A at the side of this worksheet.

STRIP A

2 Wrap it around a pop can —just like a head band— and clear-tape the ends. Keep "A" upright.

STAY BELOW THE CURVE

TAPE ONCE.

STRIP A

PULL TIGHT AND EVEN!

3 Bend the pull-tab back and forth until it breaks off.

4 Cover the hole and *half* of the center rivet with masking tape.

RIVET

Keep tape OFF THE RIM.

5 Drape thread over the top so it crosses the center rivet, and tape.

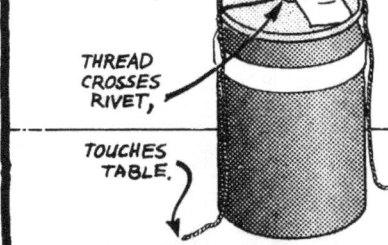

THREAD CROSSES RIVET,

TOUCHES TABLE.

6 Stick a second piece of masking tape exactly over the first.

THREAD IS "SANDWICHED" HERE

7 Tape a penny to the thread so it hangs just off the table. Trim both ends.

8 You can "dial" any measure just by pushing the thread around the rim.

TRY IT!

A COMPUTER CAN!

9 Work with a friend who also has a computer can. Follow these three steps.

1 *DIAL* each measure below.

Do 5.5 first...

2 *COMPARE* your answer with your friend's.

5.5?

CHECK!

3 *RECORD* your answer on the scale below.

a. 5.5 cm	e. 35 mm	i. .023 m	m. 83 mm
b. 11.1 cm	f. 134 mm	j. 6.4 cm	n. .117 m
c. 0.9 cm	g. .100 m	k. 0.1 cm	o. 46 mm
d. 70 mm	h. .140 m	l. 15.0 cm	SAVE YOUR CAN!

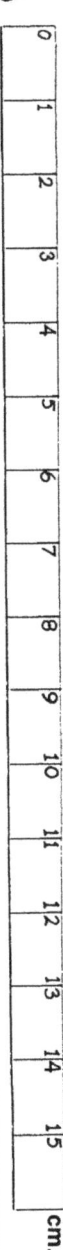

RECORD answers here:

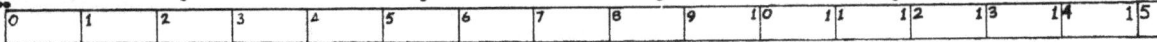

0 1 2 3 4 5 6 7 8 9 10 11 12 13 14 15

TOPS LEARNING SYSTEMS

AGREE/DISAGREE

1 Cut out strip "B" at the side of this worksheet.

2 Set the penny on top of your can, out of the way.

3 Wrap strip "B" around the can, just below A. Tape the ends together.

Tape just ONCE!

4 Find a friend to play Agree/Disagree. You'll need only one measuring can.

LET'S USE MINE...

STRIP **B**

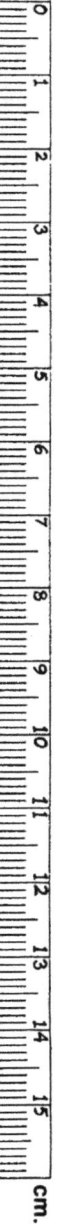

5 Play Agree/Disagree on the "A" band only. Follow these four rules.

1

SLIDE THE HAIRLINE *to any chance position.*

IGNORE "B" FOR THE FIRST GAME.

2

WRITE YOUR MEASURES *on separate paper.*

3

COMPARE YOUR ANSWER *with your friend's.*

SCORE CHART
● ● ● ● ● ● ● ●

4

MOVE A PAPER CLIP ON THIS SCORE CHART . . .

. . . UP 1 IF YOU AGREE: *Uncertain digit off by no more than one.*

. . . DOWN 2 IF YOU DISAGREE: *Uncertain digit off by more than one, or certain digits disagree.*

WIN
5
4
3
2
1
START

Mark your place with a paper clip!

6 Now play Agree/Disagree on the "B" scale only.

Ignore "A" this time.

7 Which scale is hardest to play, "A" or "B"? Why?

TOPS LEARNING SYSTEMS

LINE MEASURE

1 Carefully cut out the *cm* ruler at the bottom of this page. Measure each line, then write your answer in the box.

WRITE YOUR UNITS↓

ESTIMATE THE LAST DIGIT— DON'T ROUND IT OFF!

2 Now cut out the ruler with mm divisions. Carefully re-measure each line and write your answer in the box.

WRITE YOUR UNITS↓

a. **2.6 cm**	a.	a. **2.62 cm**
b.	b.	b.
c.	c.	c.
d.	d.	d.
e.	e.	e.
f.	f.	f.
g.	g.	g.
h.	h.	h.
i.	i.	i.
j.	j.	j.

3 Which ruler is more accurate — the centimeter one or the millimeter one? Why?

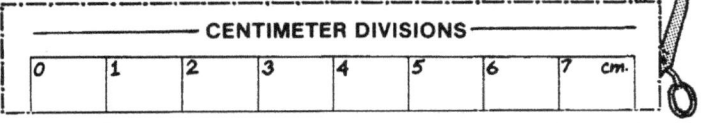

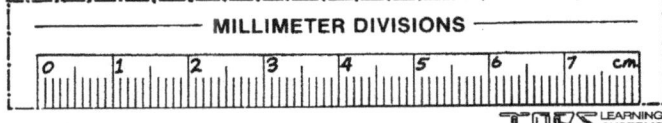

CENTIMETER DIVISIONS

0 1 2 3 4 5 6 7 cm.

MILLIMETER DIVISIONS

0 1 2 3 4 5 6 7 cm.

TOPS LEARNING SYSTEMS

METER MEASURE

1 Carefully cut around the outside of this gray box.

LEAVE THIS TAB ON.

2 Now cut your gray box into 5 long strips. Tape these strips in order — 20, 40, 60, 80 — to make a meter tape.

LINE UP WITH THE STRAIGHT EDGE OF YOUR TABLE.

3 Write your name on the back of your tape:

NAME

4 Find the length of each object in centimeters. Measure accurately — don't round off.

A 3 x 5 INDEX CARD:

diagonal:

width:

length:

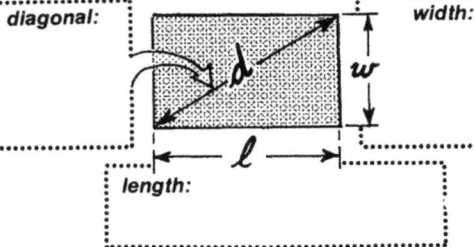

THIS WORKSHEET:

diagonal:

length:

width:

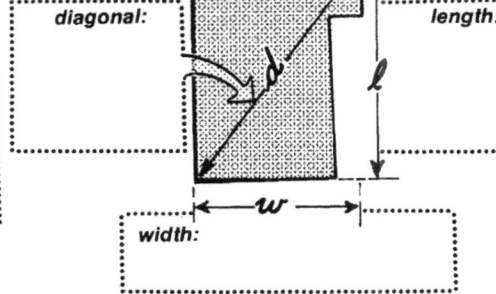

A PENNY:

width:

diameter:

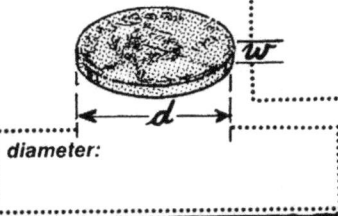

A SODA CAN:

circumference:

height:

diameter:

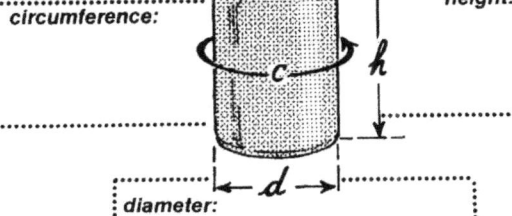

5 Underline the *certain* part of each measurement above. Check with a friend to make sure your *certain* figures agree.

CHECK...

Do all the figures in each measure agree? Explain.

SAVE YOUR METER TAPE

TOPS LEARNING SYSTEMS

BODY MEASURE

---

1 Use your meter tape to find how you size up: *Estimate* between lines, don't round off.

CIRCUMFERENCES:

Forehead: ....   Waist: ....

Neck: ....   Wrist: ....

LENGTHS:

Height: ....

Fathom: ....

Span: ....

---

2 A very famous scientist* found that 8 times your span equals both your height and your fathom: **8 spans = 1 fathom = 1 height.**

*Leonardo da Vinci

Use math to show how closely *you* fit this rule.

---

3 Is it better to measure in spans or in meters? Explain.

TOPS LEARNING SYSTEMS

BUILD A MICROBALANCE

1 Fold a 3x5 card in half the short way (like a tent). Write your name at the bottom.

2 Using a penny, draw a circle at the center of the fold, then cut a square around the circle.

DRAW CIRCLE ... CUT SQUARE

3 Unfold the square and tape it across the bottom of the tent.

4 Make a 1 cm snip in the end of a straw . . .

1 cm

. . .Squash-fit the cut end inside another straw . . .

CUT END

. . . Push them together to overlap 8 cm.

8 cm

5 Poke your double straw through the tent so the seam rests in the middle . . .

SEAM

NAME

. . . Stick a pin through both straws *exactly here.*

Near the top of the curve

1 cm from the seam.

6 Stick tape over this spoon handle where shown. Carefully cut out around the dashed line.

FIRST TAPE

THEN CUT

7 Cut to the middle dot along the dotted line. Overlap the edges so the arrows come together, then tape.

8 Curl the handle *up* so it slides into the *short* end of your straws.

SLIDE IN

ROLL UPWARD

9 Push the spoon in all the way to its bowl . . . . . . then bend it level.

PUSH IN TO BOWL

10 Fold some tape so one end is sticky . . .

HANDLE

STICKY TAB

. . . and stick it next to the spoon.

TAPE "RIDER"

11 Trim back the long end *one mm* at a time, until the spoon *just* tips down.

OK— enough!

12 Move the rider to make the beam balance level.

TOPS LEARNING SYSTEMS

NAME: _____   CLASS: _____

SEEDS & SUCH

1 Cut a straw into 3 parts like this:

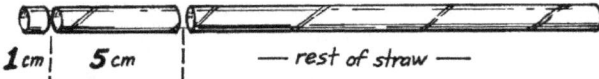

1cm | 5cm | — rest of straw —

2 Slice each section open along its entire length.

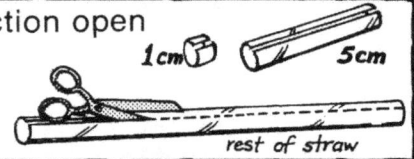

1cm   5cm

rest of straw

3 Center your microbalance: move the tape rider so the straw balances level.

MOVE THE RIDER!

4 Slide the *1 cm section* over your pencil, then onto the end of your balance.

THIS IS YOUR SLIDER

5 Put a rice grain in your weighing spoon. Move the *slider* until the beam balances.

RICE GRAIN

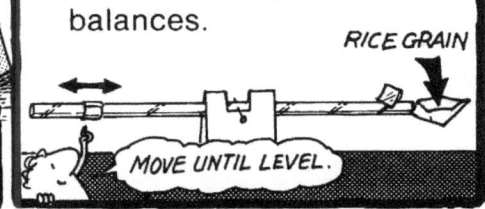

MOVE UNTIL LEVEL.

6 Does a rice grain weigh more than a staple? How does your balance tell you the answer?

7 Which is heavier, a staple or a lentil? Use your balance and straw "sliders" to find out.

LONGER SLIDERS BALANCE HEAVIER OBJECTS.

8 Use your microbalance to find the *heavier* in each pair:

CIRCLE YOUR ANSWER

pop corn / pinto bean

paper clip / pinto bean

pop corn / lentil

9 Order these from lightest to heaviest:

• staple • popcorn • paper clip •
• rice grain • lentil • pinto bean •

lighter

HEAVIEST GOES HERE

....
....
....
....
....

10 Put all 3 straw sliders in a cup.

Save them to use again.

TOPS LEARNING SYSTEMS

MILLIGRAM MASSES

1 Write your name on this gray box. Then cut it out and tape it to the side of your cup of sliders.

2 Carefully cut out each paper mass and fold it so the number shows. Store all 12 in your cup.

CUT:

FOLD: NUMBERS MUST SHOW

Name........................

This cup contains 3 SLIDERS, TWEEZERS, and 11 MASSES:

200 mg	20 mg	2 mg
200 mg	20 mg	2 mg
100 mg	10 mg	1 mg
50 mg	5 mg	+ extra squares

PAPER MASSES:

3 Make a pair of tweezers:

FINISHED TWEEZERS

Cut a straw into halves.

Bend one, and tape the elbow.

Cut the other in the middle at a slant...

Snip outside edges 1 cm.

...to make 2 POINTED ENDS.

Slide snipped ends INTO pointed pieces:

TURN ENDS SO POINTS MEET.

4 Weigh each object in mgs:

	lentil:	popcorn:	rice grain:	paper clip:	staple:	pinto bean:
MASS (mgs):						

5 Do these results agree with activity #15? Explain.

6 How many rice grains make 1 gram (1000 mg)? Show your math.

STORE SLIDERS, TWEEZERS & MASSES IN YOUR CUP

200 mg · 100 mg · 200 mg · 20 mg · 20 mg · 10 mg · 5 mg · 2 mg · 2 mg · 1 mg · EXTRA SQUARES (Don't Fold) · 50 mg

TOPS LEARNING SYSTEMS

SEED GRAPHS

1 Weigh rice grains on your microbalance. Fill in this table.

HOW MANY	MASS (mgs)
0	0
1	
4	
10	

2 Plot and circle each point on the graph. Draw the best possible straight line through your points and label it "rice grains."

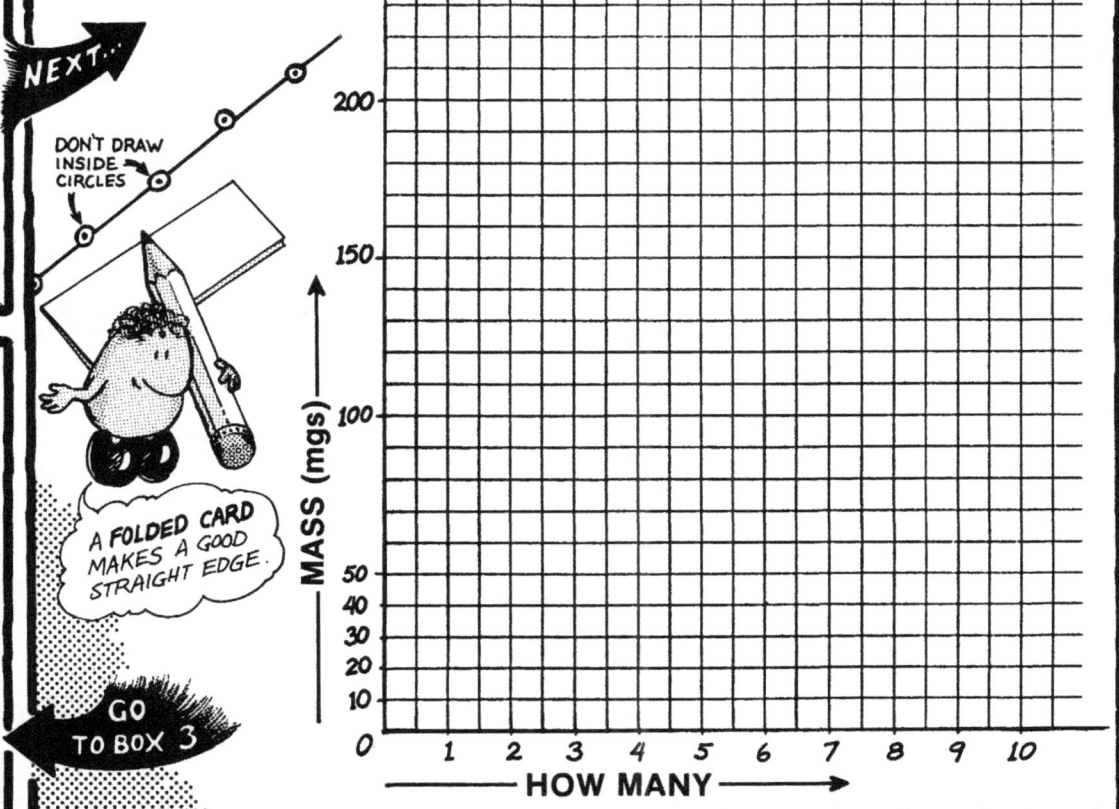

NEXT..

DON'T DRAW INSIDE CIRCLES

A FOLDED CARD MAKES A GOOD STRAIGHT EDGE.

GO TO BOX 3

MASS (mgs) / HOW MANY

3 Now fill in this table for lentils. Draw another graph line and label it "lentils."

HOW MANY	MASS (mgs)
0	
1	
2	
4	

4 Read your *graph* to estimate the mass of . . .

. . . 7 rice grains.

. . . 3 lentils.

Does your microbalance prove your graph results are correct? Explain.

CHECK YOUR ESTIMATES

5 Staples weigh about 30 mg each. Predict how a graph line for staples compares to rice and lentils.

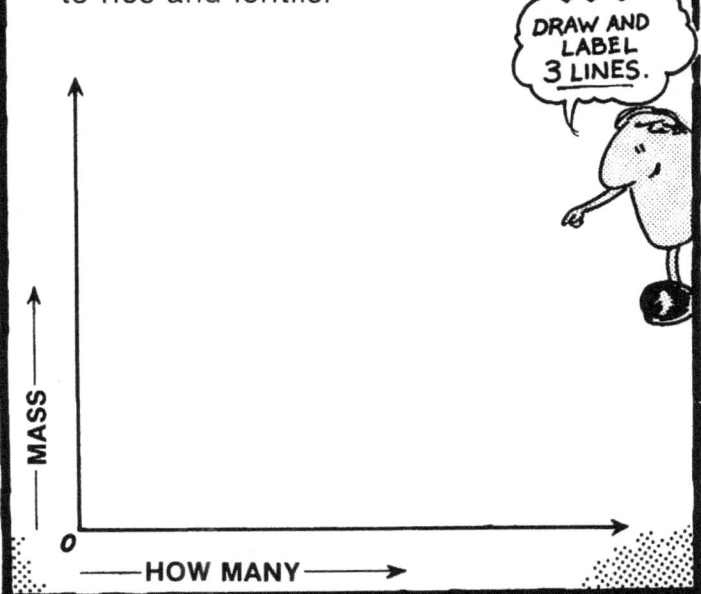

DRAW AND LABEL 3 LINES.

MASS / HOW MANY

WATER DOMES

1 Carefully cut out this "drip strip" along the dashed line.

PINHOLE

2 Tape your drip strip to a cup *exactly* as shown. The gray triangle must hang even with the bottom edge.

TAPE JUST ABOVE PINHOLE

BOTTOM EDGE

3 Cut away the top, just above the drip strip.

CUT CLEAR AROUND.

4 Cut off only *half* of the top of another cup.

AS SHORT AS THE FIRST ONE

5 Paper clip the short cup inside the tall one so the drip strip hangs in the opening.

6 Poke a pin through the drip strip hole and into the cup. Pull it back out.

7 Pour water into the top cup . . .

. . . To *start* it dripping, wet the paper on *both* sides with your fingers.

. . . To *stop* it dripping, push the pin back in the hole.

8 Balance a penny on a half clothespin. See how many drips it holds without spilling.

RECORD 10 TRIALS

CATCH DROPS IN FREE FALL. DRY PENNY AND CLOTHESPIN BETWEEN TRIALS.	1	2	3	4	5	6	7	8	9	10

9 A penny holds about 1 ml of water.

Throw out trials you think are mistakes, then find an average.

AVERAGE DROPS IN 1 ml:

10 A *level* (not heaping) teaspoon holds 5 ml.

Predict how many drops this makes:

Test your prediction:

11 Empty all water from your dripper. Put your name on it, and save it for the next activity.

TOPS LEARNING SYSTEMS

METRIC DRIPS

1 Make a micro-dish: Press a penny into aluminum foil, then cut around the outline . . . : . . Curve your "dish" by pressing it into your hand.

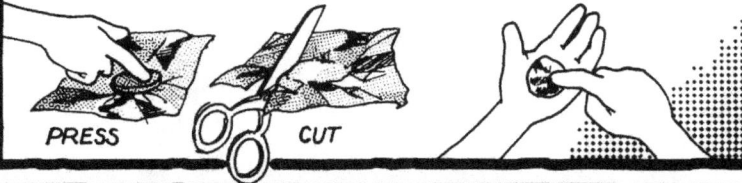

PRESS   CUT

2 Center your balance, then find the mass of your micro-dish.

HOW MANY mgs?

3 Set up your dripper as before . . . . . . Wet both sides of the paper so the drops flow *fat* and *slow* . . . . . . Catch just one on your micro-dish in free-fall.

MINE

SLOPPY WET!

MINE

Don't touch the drop before it falls.

4 Find the mass of this drop, then 2 others. Dry your dish and center your balance after each trial.

DROP:	1	2	3	
DISH & DROP				*AVERAGE:*
−DISH ONLY				
=DROP ONLY				

5 Use your result from activity 18 to estimate the total mass of water a penny can hold.

WATER DOME:

HOW MANY MGS?

6 Cover this cube pattern with tape. Cut it out and tape it together so it can hold water!

TAPE FIRST, THEN CUT OUT!

1000 mg

1 ml

1 cm.³

1 gm

7 Do you think your cube holds *more* than your penny, *less* than your penny, or about the same? Explain.

8 Test your prediction. What do you find?

...4... ...5... ...6...

TOPS LEARNING SYSTEMS

CONSUMER SCIENCE

1 Weigh uncoated aspirin tablets 3 times. Find the *average* mass.

Trial 1	.... mgs
Trial 2	
Trial 3	
Average	

2 Read the label. Notice that each tablet contains 5 "grains" of aspirin.

One "grain" equals 65 mgs. How many mgs of aspirin are in your tablet?

3 Compare: is your tablet made from *all* aspirin (100% pure) or does it contain added ingredients? Explain.

4 Read the net weight (in grams) on a small package of *unopened* candies, nuts or raisins.

NET WEIGHT means without the wrapping.

How many grams?

How many mgs?

5 Weigh just one piece on your microbalance.

You may need to make another 500 mg mass from foil.

...and a longer slider!

How many mgs?

6 Now use math to estimate how many fill the whole package. Whoever comes closest might win the whole package!

SHOW YOUR MATH!

??

TOPS LEARNING SYSTEMS

Feedback

If you enjoyed teaching TOPS please tell us so. Your praise motivates us to work hard. If you found an error or can suggest ways to improve this module, we need to hear about that too. Your criticism will help us improve our next new edition. Would you like information about our other publications? Ask us to send you our latest catalog free of charge.

For whatever reason, we'd love to hear from you. We include this self-mailer for your convenience.

Sincerely,

Ron and Peg Marson
author and illustrator

Your Message Here:

Module Title _____ Date _____

Name _____ School _____

Address _____

City _____ State _____ Zip _____

———————————————— FIRST FOLD ————————————————

———————————————— SECOND FOLD ————————————————

RETURN ADDRESS

_____

_____

_____

PLACE
STAMP
HERE

TOPS Learning Systems
342 S Plumas St
Willows, CA 95988

TAPE HERE